# Business, Marketing, and Management Principles for IT and Engineering

# Business, Marketing, and Management Principles for IT and Engineering

Dmitris N. Chorafas

CRC Press is an imprint of the
Taylor & Francis Group, an **informa** business
AN AUERBACH BOOK

CRC Press
Taylor & Francis Group
6000 Broken Sound Parkway NW, Suite 300
Boca Raton, FL 33487-2742



Printed in the United States of America on acid-free paper
Version Date: 20110429

International Standard Book Number: 978-1-4398-4806-7 (Hardback)



**Visit the Taylor & Francis Web site at**
**http://www.taylorandfrancis.com**

**and the CRC Press Web site at**
**http://www.crcpress.com**

# Contents

## Part Two Management Principles

## Part Three Marketing and Sales

## Part Four Innovation

# Preface

In a globalized economy, the management of an enterprise requires depth, horizon, and skills beyond those commonly available. This is true of all companies, and most particularly of those at the edge of technology.

Corporate growth and survival call for the appreciation of strategic crossroads and application of knowledge to reach goals that in the past were often reserved for fiction. Challenges are associated with products, processes, and markets—as well as new business opportunities and risks that go along with them.

Written for business practitioners in engineering and technology as well as for graduate students in colleges and universities, this book presents the *principles*, *policies*, and *practices of management* by the best companies, as well as the way they develop and implement them. As a rule, these firms pay a great deal of attention to matters of efficiency, productivity, and rationality that are

- Distilled into management principles, and
- Demonstrated by means of daily activities, as documented by the case studies included in this text

Case studies are of prime importance because they identify the critical issues confronting people engaged in business. They also demonstrate in a matter-of-fact way how the leaders of industry select and set strategic criteria to guide their decisions toward success undertakings.

In the background of the management principles, policies, and practices—that the reader will find organized in 16 chapters—lies the fact that no organization can survive in the longer term without being ahead of the curve. Successful managers are imaginative and flexible, adaptable to developing circumstances. Practical decisions rather than theories hold the upper ground.

***

The text divides into five parts. Part One concentrates on the important *strategic* issues guiding the mind and hand of decision makers who have the salt of the Earth. Chapter 1 discusses the components of a business strategy—from human resources to marketing, product innovation, financial, and general management. A successful strategy positions the company against market forces in a way guaranteeing its profitability and survival.

Chapters 2 and 3 address themselves, respectively, to the *principles* and *functions* of management. Management is an art, not a science; but an art, too, has its rules. These are explained by way of six most important functions: forecasting, planning, organizing, staffing, directing, and controlling. The aim is to provide leadership, which—charisma aside—is a matter of conceptual skills, analytical detail, quality of decisions, and concentration in areas of strength.

Part Two elaborates in greater depth the management function, offering to the reader plenty of case studies to better explain what is being said. The subject of Chapter 4 is *forecasting*, which primarily focuses on the future impact of current decisions, and on future events in the sense of analyzing their aftermath. Because it makes projections, considers expectations, and evaluates likely adverse forces, forecasting is a prerequisite to sound planning.

*Planning* and forecasting are twins. "The plan is nothing," Dwight Eisenhower once said, "Planning is everything." Far-out planning aims to answer the queries: What will our company be in the next 10 to 15 years? From where will its earnings come? Under the time perspective come long-range planning (5 to 7 years), medium-range planning (2 to 4 years), and the next year's plan—of which the budget is the financial framework.

*Organization* and *structure* are discussed in Chapter 5. Management's ability to run the organization is based on its span of control, span of knowledge, span of support (including IT services), and span of attention to detail. Structural choices aim to promote those abilities by delineating the line of command—describing proper relationships, establishing positional qualifications, and making available measures of performance.

The theme of Chapter 6 is that a company's most important assets are *people* and *people*—people its employees, and people its clients. Challenges include not only leading, but also recruiting qualified personnel, assigning competent people to each position, countering human obsolescence through lifelong training, and steadily improving productivity—which is a cornerstone issue in this chapter.

Chapter 7 provides the reader with the sense of *management control*—most particularly internal control. Management control ensures adherence to plans by watching over progress toward objectives commensurate with goals and time plans being established. Therefore, a basic prerequisite is the existence of financial plans, market plans, product plans, and human resources plans. Successful management control always leads to corrective action.

*Marketing and sales* are the overriding subjects of Part Three. As Chapter 8 points out, the first decision in establishing a marketing strategy focuses on the specific market(s) to which the company wants to appeal. The next is the building of human resources able to be in charge. The third is the share of the market and market rank the company wishes to achieve. Should we aim to be No. 1 or No. 2? Such decisions condition innovation, product planning, sales network, and the ability of the firm's salespeople to channel its products to the market.

The market's conquest is decided through actions, not through words. The Bloomberg, ITT, Cisco, and Microsoft case studies in Chapter 9 provide irrefutable evidence to this fact. Imaginative ideas that have not yet been tried by competitors can give *our* company a head start, but the market's conquest must be approached through a coordinated program, and this is provided by the marketing plan.

Chapter 10 concentrates on sales tactics and on sales proper. The marketing functions must be detailed by means of practical steps promoting the sales effort. Sales efficiency and control of the sales network's deliverables are part of these steps. In addition, a properly trained sales force is a valuable asset.

Part Four centers on product development and innovation. Chapter 11 explains what is meant by *technology* as a generic term. No innovation nor no new process is anything if it does not create value to the company and to its customers. Quality, added value, and cost effectiveness are cornerstones all the way from product planning to marketing decisions. They are also highly dependent on how management links technology to markets.

Chapter 12 concentrates on *product planning* and *pricing*—which is never made in the abstract but reflects a pragmatic view of product features, customer drives, market dynamics, and competition. To a substantial extent, product pricing is conditioned by whether a company is a leader or follower in its market, as well as by the nature of this market: mass, unique product, or niche. Product pricing is also dependent on the cost structure decided at the time of product design and upheld during production, distribution, and maintenance.

*Price wars* in the computer industry, a wholesome case study, is the subject of Chapter 13. As far as computers, their makers, and their customers are concerned, the price wars of late 1970s have been a strategic inflection point, akin to the break-up of AT&T by the decision of Judge Green in the 1980s. Until then, IBM was by far the dominant decider of computer design and of prices. Other computer companies read IBM's price list and tried to do a little better.

But the late 1970s computer industry price wars turned this comfortable price leadership on its head, unleashing a wave of competition. This has been followed by a forward leap in technology and by the switch to Vaxes, PCs, LANs, client-servers, database machines, supercomputers at affordable cost and, more recently, cloud computing.

The theme of Part Five is the financial aspects of enterprising. Chapter 14 concentrates on *financial management*, and most particularly on budgeting: the short-term financial plan. Contrary to what is generally thought, a budget is no authorization to spend money. It is a plan matching spending needs—including operational costs and investments—with financial resources. Budgets must be flexible and adaptable, alternative budgets being one option; and they should be carefully controlled.

A company's *cash flow*, the other subject of Chapter 14 comes from sale of its products and services, depreciation, amortization, retained earnings, and sale of assets. A sustainable cash flow commensurate to the company's assets and plans of operation underpins financial staying power. Cash flows are reduced by servicing debt that raises the argument of management's equity versus debt decisions.

*Costs* matter. Chapter 15 is dedicated to profit centers and cost control. Profit centers are income earners; cost centers survive through budgetary allocations. Therefore, organizational structure based on the former is the preferred solution. A neat profit center organization must be supported through standard costs, with plenty of attention paid to effective control over expenditures.

Chapter 16 concludes this book by bringing the reader's attention to longer-range financial planning, management accounting, and reliable financial reporting. The most advanced information technology should be used to promote management's ability to be in charge of financial matters. Virtual balance sheets are an example.

Successful management is never based on vagueness. The 16 themes treated in an equal number of chapters and their case studies provide evidence to this statement. Only profitable firms can survive in the longer run, but profitability can never be taken for granted. Even the best analysis of future income can be misleading because of inaccurate, delayed, or outright false financial data, or by drift often beset by management whose time is almost up.

***

In each of the themes to which it addresses itself, the text examines the principles of management and how they can be put into practice in a pragmatic way. It also brings to the reader's attention the positive and negative aspects of different policies and whether or not current practices related to forecasting, planning, organizing, staffing, directing, and controlling have been producing the required synergy.

# Acknowledgments

I am indebted to a long list of knowledgeable people and organizations for their contributions to the research that made this text feasible. I am also indebted to several senior executives and experts for constructive criticism during the preparation of the manuscript, most particularly to Dr. Heinrich Steinmann and Dr. Nelson Mohler.

Let me take this opportunity to thank John Wyzalek, for suggesting this project, Kyle S. Meyer, Jr. for seeing it all the way to publication, and Papitha Ramesh for the editing work. To Eva-Maria Binder goes the credit for compiling the research results, typing the text, and making the camera-ready artwork and index.

# Part One

# Business Strategy

# 1

# *Strategy*

## 1.1 BUSINESS STRATEGY DEFINED

By *strategy* we understand the art of conquest by means of a master plan. A strategy may be personal, corporate, or national, or it may be civilian, cultural, religious, or military. Buddha said, "Between him who conquered in war millions of men, and him who conquered himself, the greater victor is the latter." The conquest of oneself is the supreme strategy.

Some strategies prove to be brilliant; others are dumb or outright disastrous. The error, however, is not always in the master plan. "The value of a strategy is that of the people who apply it," said Jack Welch, the former chief executive officer (CEO) of General Electric (GE). His dictum applies in the dual sense of

- successes, and
- failures

"Once we have learned our lesson from a failure, we must again assume the risk commensurate to the job we are doing," Walt Disney once suggested. In the opinion of Bernard Arnault, of LVMH and Christian Dior, "errors are inevitable—but properly used they can serve in forming one's mind." What should be avoided is falling into the abyss. Therefore the master plan should be dynamic, adaptable, with a "Plan B" for fallback.

Business strategy achieves its aims through a plan (Chapter 4) consisting of tactical moves that concern the execution of operations under the master plan. *Tactics* are a vital ingredient of strategy, but they are subordinated to it. The four fundamental components of any master plan are as follows:

1. *Human Resources Strategy* (Chapters 2 and 6)—Aiming to identify, recruit, and keep developing the company's managers and professionals.

Andrew Carnegie was once asked if he were to lose all his factories but keep his personnel, or lose all his personnel but keep his factories, which option would he choose. He answered the former, because if he had the people he selected and trained over the years he would rebuild his empire. A company's greatest asset is not the money in its vaults but *people* and *people*: its customers and its employees.

2. *Financial Strategy* (Chapter 14)—Focusing on forward planning to protect and grow the entity's economic resources in spite of uncertainties and turbulence.

An effective financial strategy requires positioning the company against market forces. To do so, board members, CEOs, chief financial officers (CFOs), and their colleagues must raise their eyes above the day-to-day commitments and face their broader responsibilities of safeguarding and growing the company's assets.

3. *Market Strategy* (Chapter 8)—Focusing on effectively identifying the company's market, managing the change occurring in the market, and developing a plan able to sustain leadership.

The more dynamic the market is, the greater the need for market repositioning. Of the top 100 American industrial companies ranked by assets in 1914 when World War I started, only one has been in this select list in 2010: GE. Most of the others disappeared. Of the top 100 in 1945, at the end of World War II, only a handful are top ranking these days, and the then No. 1, General Motors (GM), had to go into bankruptcy to restructure itself and try to survive.

4. *Product Strategy* (Chapter 12)—Aiming to be in charge of products and services offered to the market, and pricing them in a competitive but profitable way to confront the moves made by our company's challengers.

With fast advancing technology, what companies need to position themselves into the market is to not only gain recognition but also retain their brand's attractiveness. This calls for innovating and developing products the rest of the industry does not have. Such products however, should fit within the company's strategic thinking.

No entity can be a leader in all fields, but it must lead at least in some products that will become its key market thrust and income earner. It should be always remembered that strategic products change over time because of moves by competitors and the market's inflection points that bring companies at crossroads (Section 1.3).

As these references document, business strategy cannot be made in the abstract. It must be factual, and it calls for exercise of skill, forethought, and artifice in reaching goals and carrying out plans. Strategy is a dynamic process characterizing the interaction of two or more persons or entities each of whose actions is based on certain expectations of moves by the other, over which the first has no control. This definition contains the three elements that set strategy apart from other plans or moves:

- It is a *master plan*, not just a list of individual actions.
- It is *against* hence it involves competition, a basic ingredient in any strategy.
- It has an *opponent*, another person, group, organization, or nation, without which the competitive situation could not exist.

In the sense outlined by these three bullets, strategy has been applied to war, business, and propaganda, as well as politics. Successful strategies follow an ancient military maxim that states, "Reinforce success, abandon failure." Strategy is not an objective in itself, but a master plan to accomplishing objectives.

One of the fundamental notions underpinning a master plan is that of identifying our side's strengths and weaknesses, putting forward moves that reinforce our strengths and estrange our weaknesses. In his book *Competitive Strategy*, Michael Porter, of Harvard University, asked how companies can secure long-term competitive advantage in the realm of two conflicting schools:

1. The Harvard Business School, which urged companies to adjust to their unique circumstances
2. The Boston Consulting Group, which argued that planning can be based on a "universal principle," the so-called experience curve

According to the second hypothesis, the more a firm knows about a market, the more it can lower its price and increase its market share. Porter tried to find a middle way by studying individual companies and setting them in the context of their industry. From this he deducted *generic strategies.**

To my book, "generic strategies" are the wrong approach, and "universal principles" should be used only by the unable who has been asked by the unwilling to do the unnecessary. Although one can learn a great deal by reading Sun Tzu and Machiavelli (among other great thinkers—see Section 1.2), a successful master plan is not made of prefabricated components. It is a work of art based on the strategist's

- imagination,
- training,
- experience,
- information, and
- thinking

A successful strategy is specific and has locality. It can be formulated only after the *objectives* of a person (enterprise, group, the society) have been established. The making of strategy involves projection into the future and selection among alternative plans of action. Projections and forecasts help reduce but don't do away with the effects of the unknown. Selection among alternatives is the basic process that we use to optimize a chosen course of action.

- No organization can lead its way within a competitive environment in an effective manner, without some form of formal or informal strategy.
- Reasons generic to the enterprise and its environment not only require strategic reevaluations but also call for corrective actions and reformulation of strategy.

For every form of organization—state, community, enterprise—strategies should not be strict, inflexible, monolithic, or obsolete. The existence of an established strategy should never be thought to allow the chief executive and his immediate assistants to dissociate themselves from an active

* Though also stating that different firms must choose different paths to success.

follow-up of the affairs of the organization on the assumption that, having given orders to execute everything will fall into place.

President Truman once remarked that of all the challenges and headaches of the presidency—the Cold War, the Berlin crisis, Korean War, MacArthur's confrontation, the recession—there was not a greater one than giving orders, expecting them to be followed, and then seeing that nothing happens.

The formulation of a winning strategy—industrial, financial, military, political, individual—involves the reevaluation of one's own plans, educated guesses on those of the adversary, imaginative moves to match them, and a great deal of supervision and follow-up. Control action (Chapter 7) is necessary to make sure that a plan, once established, is put into effect.

## 1.2 SUN TZU AND MACHIAVELLI

Two thousand five hundred years ago, Sun Tzu, the great Chinese statesman and general, said, "If you know yourself and know your enemy, you will not have to fear the outcome of one hundred battles." But then he added, "To win one hundred victories is not the acme of skill. To subdue the enemy without fighting is the supreme excellence."* In Sun Tzu's opinion,

- every person's invincibility depends on oneself,
- but the enemy's vulnerability depends on the enemy himself

The translation of Sun Tzu's principles of strategic moves into a business strategy ranges from building up human resources to financial planning and the tuning of the firm's activities in response to shifting market demand. The critical element will be overtaking, not just counteracting to, the actions of competitors. Therefore the crucial questions management should ask itself are as follows:

- What's *our* strategy?
- What's the strategy of *our* opponents?
- Can the strategy we have chosen leave them behind?

* Sun Tzu, *L'Art de la Guerre*, Flammarion, Paris, 1972.

The next critical queries senior management should ask itself and ask for clear answers are as follows: Which of our products are losing market appeal? Why? What are their sales? Profit margins? Costs? Risks? How do our products hold against those of our competitors? How are our clients and competitors looking at our strengths and weaknesses? Are they taking advantage of the latter?

When we talk of products and services we also talk about markets. Is the primary market to which we appeal changing? The secondary markets? How do we stand in our primary market against our competitors? How fast does this market grow? How fast does it shift? In which direction is it shifting? Which are the special risks the market's move presents to our firm's future?

Borrowing a page out of Sun Tzu's book, these questions have to be answered in a factual and documented manner prior to even thinking about possible moves. The answers to be provided will talk a great lot about business opportunity and pending disaster; and about how well our organization has positioned itself against market forces—which essentially means about order and disorder in our enterprise management.

To Sun Tzu's book, which has looked at this issue, order and disorder are conditioned by structure, organization (Chapter 3) and direction (Chapter 6). Strength or weakness greatly depend on tactical disposition, whereas courage or cowardice depend on circumstances. Heroes and cowards are the same people, Napoleon once said. Which comes up depends on the way the battle goes. Sun Tzu believed that a skilled commander seeks victory from any situation he confronts. He is the first to assume responsibility for decisions rather than relegating it to his subordinates.* And although rewards are part of the game, too frequent rewards indicate that the man in command is at the end of his resources—too frequent punishments prove that he is in acute distress.

Another lesson to be learned from a successful commander like Sun Tzu is that the best policy is to attack the enemy's strategy. All warfare is based on deception. Therefore, when capable of attacking, feign incapacity; when active in moving troops, feign inactivity. In addition,

- if your opponent is of choleric temper, try to irritate him, and
- if he is arrogant, try to encourage his egotism

* Idem.

The great Chinese general also states that it is much more complex to devise a strategy—the master plan—than to lay a plan for a battle. The battle plan is relatively short term, whereas a strategy and counterstrategy have to be devised with both the fundamental and long-term interests in mind. Expediency is no way to plan for success, and it can lead to disaster.

- The choice of the primary factor is most decisive.
- One should never be misled by complex minor factors, which tend to accompany a goal.

In the opinion of Nicolo Machiavelli, "Nothing makes a prince so much esteemed as great enterprises and setting a fine example. (But) never let any administration imagine that it can choose a perfectly safe course." "Prudence," says Machiavelli, consists of

- knowing how to distinguish the character of troubles, and
- the ability of choosing the lesser evil

The "lesser evil" is, for example, what was selected by Kennedy when, during the Cuban missile crisis of 1962, he had a group of 15 advisors set out alternatives, which were not necessarily mutually exclusive:

1. Do nothing.
2. Use warning, diplomatic pressure, and bargaining.
3. Try to split Castro from the missiles.
4. Set up a blockade.
5. Order an air strike against the missiles.
6. Undertake an invasion.

Prudence suggested alternative No. 4 as the lesser evil. Once this choice was done, there was no let-up. A plan of action assured continuity and, eventually, success in avoiding a nuclear confrontation. Alternative No. 4 led to favorable terms of a settlement, while an invasion of Cuba might have led to nuclear holocaust.

Once the lesser evil was chosen, Kennedy applied Machiavelli's way of looking at leadership in a confrontation. Actions must arise one out of the other in such a way that men are never given time to work steadily against the master plan. A businessman who put Machiavelli's advice in practice has been Charles Revson, the architect of Revlon. He is quoted as having boasted, "I built this business by being a bastard, I run it by being a bastard. I'll always be a bastard."*

"Be nice, feel guilty, and play safe; if there is ever a prescription for producing a dismal future, that has to be it," said Walter B. Wriston, the former chairman of Citibank. The aphorism applies hand-in-glove to the avoidance of tough decisions by leadership, be it at the corporate, national, or international level—in business or in politics.

Neither Sun Tzu nor Machiavelli had a stomach for shared leadership in war and politics—and by extension in the corporate world. Business history is littered with examples of companies such as EADS (the European defense group), Unilever (the Anglo-Dutch consumer goods company), and Goldman Sachs (the American investment bank), which have, at one time or another, had two captains at the helm. Almost all of these relationships have ultimately come unstuck.

Successful technology firms such as Apple and Oracle are run by their CEOs, not by split management or by committees. "A committee cannot run a company," Henry Ford once said, "just like a committee cannot run a car." Both Apple and Oracle have shone during the downturn and thrived despite the sweep and pace of globalization. Their bosses, Steve Jobs at Apple and Larry Ellison at Oracle, are by no means consultative, sharing types.

It is no less true, however, that in what sometimes appears to be a split management responsibility, where joint bosses keep a wary eye on one another, someone else may be steering the tandem. In his book *Caveat*, Alexander Meigs Haig, former Secretary of State and NATO commander, put it this way: "The [Reagan] White House was as mysterious as a ghost ship; you heard the creak of the rigging and the groan of the timbers and sometimes even glimpsed the crew on deck. But which of the crew had the helm? … It was impossible to know … ."

## 1.3 STRATEGIC CROSSROADS

There is a general recognition among well-managed companies that they cannot operate successfully unless they have a master plan, which they

* *The Economist*, March 20, 2010.

dynamically update. Such a plan helps them and guides them on how to be in charge of market developments—a reference valid about goals to be reached, and in regard to moves to be followed against competition.

Forecasts, plans, policies, procedures, and basic decisions are frequently made within the realm of a market that evolves steadily. Their most frequent downside is that they presume this evolution happens in a relatively linear way. This is false, because even if the way is sometimes linear, it is often upset by dynamic market developments, creating *inflection points*, which bring the company and its senior management at strategic crossroads.

A master plan that follows Sun Tzu's advice (Section 1.2) is one helping the organization to take a proactive role. This typically requires that potential major problems are avoided by attacking their roots early enough. The challenge in doing so lies in the fact that management decisions

- are made in the face of uncertainty, and
- are usually taken with insufficient information

What distinguishes the leaders of industry from common folk in management positions is that, in spite of these two bullets, they see the major problem coming and defuse it when it is still time to do so. In the week of April 12, 2010, Ben Bernanke, the Fed chairman, deposed to the U.S. Senate Banking Committee. To the question of a senator concerning whether he saw the big financial bubble coming, which led to the 2007–2011 deep economic crisis, he answered that he did not.

In other terms, this major inflection point in the U.S. economy, the global economy, and, in particular, the banking industry had escaped Bernanke's attention.* The next day, Chuck Prince and Robert Rubin,† the CEO and chairman of the Executive Committee, respectively of the badly mismanaged Citigroup, deposed to the U.S. Senate Banking Committee. They, too, said they did not see the big financial bubble coming.

A day later, that same week in an interview by Bloomberg Financial News, George Soros, the global investor, said both he and Paul Volker (the former chairman of the Fed) saw the big bubble coming, and he could not

* He was, at the time, vice chairman of the Federal Reserve System.

† Rubin was also Secretary of the Treasury under Clinton and a former CEO of Goldman Sachs.

understand why Bernanke, Prince, and Rubin did not. Asked about what did he do when he saw the bubble Soros said that

- he joined it because that's his way of making money,
- but, by contrast, it was the job of the Fed and other regulators to defuse the bubble before deeply damaging the American economy

Seeing ahead of time that major potential problems are brewing up and defusing them before they become king-size is one of strategic management's most important duties. How crucial this is is best appreciated by keeping in mind that the board, CEO, and senior management deal with the allocation of finite resources, some of which may be already committed when the inflection point alters the direction of the market and of the industry.

A *strategic inflection point* is a time in the life of a society, an economy, a company, or a person when the prevailing fundamentals are about to change. The good news is that it provides decision makers with the opportunity to break out of the status quo and thrust into a higher level of achievement. The bad news is that it can be deadly if

- it is not attended to in time, and
- it is not being handled in an able manner

A strategic inflection point puts under stress the board's and CEO's ability to be in charge, because it brings the company at crossroads and allows no delays or slippages in decisions. Winners and losers will be divided by their ability to steer to the right direction, while continuing to add value to products and services provided to customers.

This has happened, for example, with the deregulation of the financial and banking industry in the 1970s and 1980s. Some banks rose to the occasion, whereas others just stood by and watched their primary income sources disappear while their customers became their competitors. The lesson learned from that experience was that

- *if* top management fails to take control and steer the company in a new direction while there is still time,
- *then* its customers, competitors, and shrinking profit margins will make the decision instead of the board

The first larger issue is the ability to foresee the coming change—the choice of a new strategic direction, putting in motion associated major cultural changes, which quite often bring along internal stress and crisis. Ironically, the more successful one has been in the old structure, the more difficult it is to make the switch.

In his book *Only the Paranoid Survive*, Andrew Grove provides excellent examples for the reasons why the CEO must have not only a clear vision of the coming business landscape but also the determination necessary to lead a group of people through a tough set of changes. Building on his experience at Intel, Grove says that "It takes objectivity, the willingness to act on your convictions and the passion to mobilize people into supporting those convictions. This sounds like a tall order, and it is."*

When Intel and its CEO confronted the strategic inflection point to which Grove makes reference, the company was profitable and highly focused on tangible results. The challenge can best be described by what Grove calls the *10X* technology factor, an order of magnitude change, which did not come as a big bang but approached "on little cat feet."

The "little cat feet" were Japanese. Their owners first encroached into Intel's primary market, then advanced their lead. Following that, it looked as if they were positioning themselves to sweep it clean. Intel's position was threatened, but the strategic inflection point that confronted it also offered opportunities, not only threats. The way Grove relates it, in the mid-1980s

- The situation was "adapt or die."
- The strategic decision for adaptation has been to focus on leading in the market in microprocessors, leaving the former fief of semiconductors to competition.

*Timing* counts for a great deal. Actions associated to a strategic decision taken early on may well fall short at a later day because they would not be enough to turn the situation around. Moreover, when top management is meandering, the company gets demoralized, the people lose their direction and nothing really works.

Knowing where the organization is headed is a good incentive to employees to offer a better performance. Management's task, however, continues being complex because major choices of direction associated to a strategic

* Andrew S. Grove, *Only the Paranoid Survive*, Currency Doubleday, New York, 1996.

inflection point require depth and breadth of knowledge, as Figure 1.1 suggests. Also the guts to take tough decisions and see them through.

Sometimes, as Sun Tzu aptly suggests, a well-orchestrated deception or corporate opponent provides the company that made up its mind with the time it needs to be ahead of the curve in its new direction. World War II provides one of the best examples of throwing the enemy off base.

The case to which I am referring dates back to August 1940. The Luftwaffe had put a concerted effort to knock out the Royal Air Force's (RAF's) chain home radar station at Ventnor, on the Isle of Wight. German planes bombed the antenna and the transmitter shacks, damaging them badly but with heavy losses in men and material.

The German command was not sure if the damage inflicted to the enemy was worth the several planes downed by the British. Therefore, the next day they sent reconnaissance aircraft, and their crude detectors showed that Ventnor was back on the air. Goering decided that if it was that hard to knock out the radar site, it wasn't worth it, and the Luftwaffe stopped trying.

This had been a well-orchestrated deception. The British radar wasn't really working, but the technicians refused to admit that it had been knocked out. So they put some radar-frequency noise and oscillations into an amplifier and pretended that it was a radar transmission. The bluff worked, as some bluffs do.

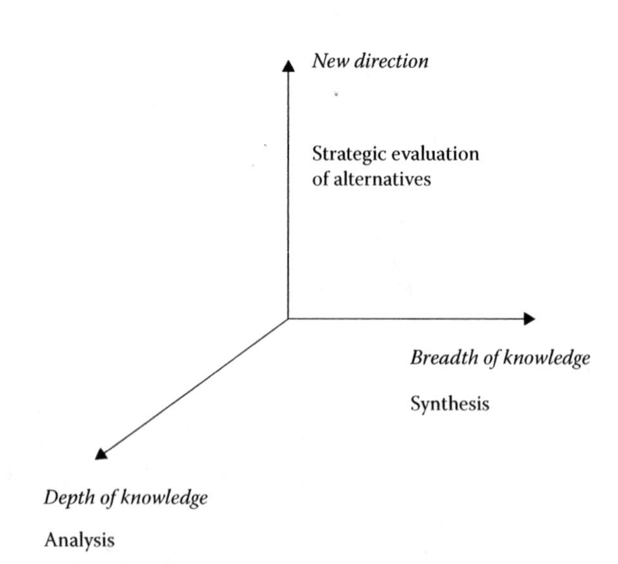


**FIGURE 1.1**
Strategic choices of direction, associated to a strategic inflection point, require both depth and breadth of knowledge.

## 1.4 EXAMINING CAUSE AND EFFECT

An in-depth analysis of past and current challenges, as well as of decisions and their aftereffect, provides plenty of insight on events, which might have added up to a strategic inflection point. Walkthroughs involving past decisions on major events are also an excellent policy for judging how effective major management moves "have been." Simulations, "what if" analyses, and playbacks offer other dimensions.

This evidently requires a wealth of data on decisions and their aftereffect(s). Learning from what was at the time a new Pentagon policy for budget overruns, in the late 1980s, several companies developed what became known as Corporate Memory Facilities (CMFs). One of the first examples has been Bankers Trust's Magellan, designed with the objective to

- register the elements of decisions and arguments that led to those decisions, and
- facilitate walkthroughs examining how sound these decisions were and what was the role played by their makers

The concept underlying this approach is *cause and effect analysis*. Virgil, the Roman poet, author, and philosopher said, "You must know the causes to draw the right conclusions." For instance, in terms of causes, the 2007–2010 deep financial and economic crisis we have been going through is due to

1. massive increase in debt, at all levels of society
2. a shadow banking system, with its zooming debt and funding risks
3. limitations of financial engineering and model-based estimates
4. excessive risk, particularly with exposures risk-takers don't understand, and
5. totally substandard risk management, as well as supervisory practices by government

It is more or less common knowledge that today in the United States (and to a lesser extent Europe) to every $1 in production corresponds $4 in debt. Yet, no government, central bank, or supervisory authority has been moving to correct the balances. Laxity has led to superficial lending practices, a real estate bubble, and the abyss of subprime mortgages—while ultraleveraged big banks disintegrated like houses of cards and had to be (unwisely) rescued with trillions of dollars of taxpayers' money.

Good managers, top scientists, and other first-class professionals appreciate the need to study cause and effect and to derive appropriate conclusions. In his *Essai-Philosophique sur les Probabilités*, Pierre Simon de Laplace, the renowned French scientist of the late eighteenth century, stated that present events are connected to preceding ones by a tie based on the evident principle that a thing cannot occur without a cause that produces it.

In a similar way, future events are to a significant extent based on present events and decisions, as well as their aftereffects. Speaking from personal experience as consultant to a diverse group of industrial and financial organizations, many critical decisions are taken in a way unstuck from the company's primary business. Therefore, as a matter of policy, Peter Drucker has been asking the businessmen he met

> What business are you in?

To this, he did not expect an answer consisting of a couple of words, but a causal reference made by the other person about his activities at a senior management position. This pushed the counterparty to critically examine his or her job. The next question Drucker asked was

> What do you need to do your business better?

To give a factual and documented answer, the counterparty had first to translate to himself or herself the information about his company's strengths and weaknesses and the causes behind them. Then, he or she had to express that conclusion in a comprehensible way, which included the causes. But Drucker pressed on

> Where do you expect to be in a year's time?

Drucker reinvented the Socratic approach, and he did it in a way obliging the counterparty to think ahead. The transition for cause to effect, and then to a new cause, is instrumental in uncovering interesting human behavioral factors. It also helps to demonstrate whether a person has thought about the past, present, and future.

Investigative work is done in a similar way. In statistical analysis we do a test of causality by examining cause and effect underpinning time series. Then, after having established the prevailing pattern we

- elaborate hypotheses taken as proxies for causes, and
- project on likely aftermath, which is taken as effect(s)

In the late 1990s, when he was treasurer of Barclays Bank U.K., Dr. Brandon Davies, who had started his career as a rocket scientist, asked his assistants to analyze the behavior of the British economy over nearly 180 years, starting with the end of the Napoleonic wars.

What Davies was after was the patterns characterizing these nearly 180 years, the existence of outliers, their causes and their effects. This put him in a position to project on future events through analogical thinking. Such an exercise is one of the best procedures available at the junction of qualitative and quantitative analysis, where cause and effect plays a major role.

The reader should know that a fairly similar approach has been used by the comparative advantage theory, developed by economists in the nineteenth century. Its primary objective was to understand and explain industrial location. Examples of causes are as follows:

- Possession of natural resources
- Skilled labor being required
- Amount of capital per worker
- Particular climatic conditions

Comparative advantages impact most significantly on strategic decisions. Cotton was grown in the south of the United States because the climate and soil were right. But it was woven into cloth in New England because that's where water power and local skills existed. Similarly, steel was made in Pittsburgh since this minimized transportation costs given the waterways connecting to iron ore and coal deposits.

The mobility of material assets is limited by costs. Transporting cement by rail for more than 200 km is uneconomic. By contrast, knowledge resources are very mobile. Brains go where they are appreciated, Robert McNamara said in the 1960s at the high water mark of the U.S. brain gain.

The knowledge industry is not the only reason why past notions concerning links existing between cause and effect have changed. Over the last decades, deregulation, globalization, and innovation altered the economic laws governing capital and labor, as effectively everyone borrows, and big money is lent in practically any major financial center.

- Financial capital is widely accessible at a price.
- Human capital can be created nearly anywhere through a consistent educational effort, at least among the intellectual elite.

The emphasis on knowledge and skill is both the cause and the result of the ongoing global change, with the effect that the raw form of unskilled or barely educated workers is less and less needed. At the same time, however, there is a fast growing obsolescence of skills, whose cause can be found in rapid technological advancement.

Through experimentation, simulation, and modeling the sciences, too, have gained a new insight into cause and effect relationships. Key to this is the analytical evaluation of quantification's results. For instance, one way to test mathematical models for accuracy and resilience is through what is known as "forecasting the past," by

- turning back in time the series
- taking a tranche, for instance, 10 years
- applying the model being developed or tested to a period prior to that tranche (if necessary adjusting the parameters), and
- comparing the emulated results with real-life statistics

Plenty of evidence can be provided by examining how close the model's output approximates statistics and other data from the years following the tranche, which has been used as the basis for modeling. A sophisticated mathematical artifact would be adjusted to reflect causes from the timeframe it is supposed to pertain, in order to provide effects comparable to those of real-life statistics from the period after the tranche.

## 1.5 SALIENT ISSUES IN INDUSTRIAL STRATEGY

Cause and effect analyses have been effectively used in the study of salient problems confronting industrial strategy. A *salient* problem is the one to which management must immediately address its undivided attention. A company, like a person, may have many problems, but at any given time only one of them is salient.

By majority, though not exclusively, the salient problem of the moment, which confronts an industrial firm will have to do with one of five areas: innovation, quality, cost production, distribution—or, sometimes with a combination of them. Intentionally, a reference to financial problems has not been included in this short list, because in the majority of cases they are the effect of other problems—not their origin.

In a financial seminar featuring case studies analysis in 1954 at UCLA, Dr. Louis Sorel and Dr. Neil Jacoby, the Dean of the School of Business Administration, had forbidden their students from suggesting financial problems as salient. They did so for the very reason mentioned in the preceding paragraph.

By contrast, salient problems have much to do with the product line—and this quite frequently. Indeed, innovation is not simply a matter of new products and services. The concept is much broader, and relates to the ability of a company to reinvent itself.

- Practically every industry is reaching a point where the next step needed to ensure its future challenges management's ability to make up its mind.
- New ideas and redefinition of the business do not guarantee survival, but lack of them increases tremendously business risk.

In the 1990s when Silicon Valley was making global strides the saying was that its companies, which wanted to continue being in business had to reinvent themselves every 1½ years. That way of thinking was based on the fact that companies assume enormous business risk by falling behind in innovation and market appeal. Sometimes this lag is compounded by management risk, and it can get amplified by exposure in an industry that is in decline.

Business risk, management risk, and industry risk correlate. As Figure 1.2 shows, at the core of their synergy is exposure resulting from design delays and failures, engineering "snafus," substandard marketing policies, high labor costs, high overhead and lightly assumed commitments. Commitments made by big companies in health plans and pensions provide an example.

In the early years of the twenty-first century, GM reported annual losses of $24.5 billion. Of this wholesome amount only $1 billion came from operating losses. The origin of the lion's share were health care commitments by GM, lightheartedly made by management as (under the old GAAP accounting) they did not show in the balance sheet. But with the newly enacted GAAP accounting standards,

- they had to be written down, and
- this led to a torrent of red ink

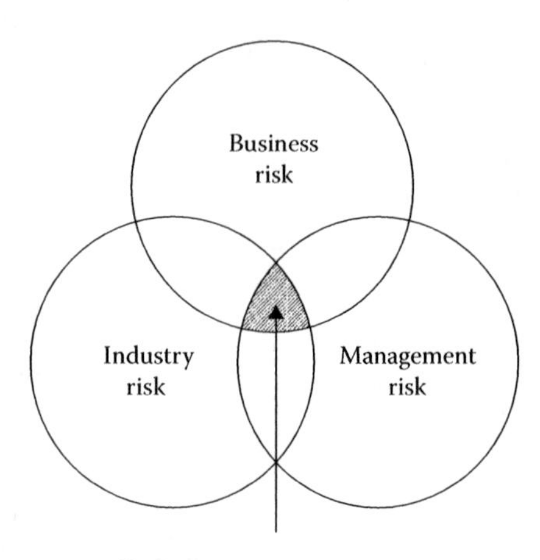


**FIGURE 1.2**
Business risk can be found at the junction of industry risk, business risk and management risk.

Much more frequent in industrial decisions are errors made in the engineering, production, and marketing domains. Hence, the importance to examine in a fundamental way the real significance of product strategy by asking critical questions with a bearing well into the future:

- What will our company do in 20 years?
- From where will come business and profits?
- How sustainable will they be? What needs to be done to protect the customer base?

Answers to these queries often require reexamining strategic decisions. Should our company be a product leader in every field of its endeavor? Should it develop all products through own R&D? Buy patents? Proceed through acquisitions?

Other queries asking for a soul-searching response are as follows: Is there enough management skill to handle the new products? To open new markets? Which are, for *our* firm, the negative effects of globalization? The opportunities? How can we overcome the negatives? What do we need to amplify the business opportunities?

Here is how Amazon.com approached this subject. To wave any ambiguity, Jeff Bezos said since day 1 that Amazon.com aims to be a customer-centric entity where people and companies can find anything they may want to buy online. Hence, it will continue to expand the range of products and services offered to its customers and partners using technology to

- increase efficiency, and
- reduce cost, to the benefit of its clients and of itself

Amazon.com's platform consists of global brand recognition, the firm features a growing customer base, and its management is known to spouse innovative technology to provide fairly extensive fulfillment capabilities. Over the years, the company built a most significant e-commerce expertise, having worked hard with the aim to

- launch new Internet commerce businesses quickly, and
- endow it with good prospects for market success

Having established an infrastructure for the products it offered to the market, Amazon's senior management attacked the next salient issue: how to use this infrastructure to develop and market new products and services. Behind this strategic move has been the will to better amortize the infrastructure and expand the company's income stream. The answer has been *Cloud computing.* Today Amazon.com is one of the major cloud computing players worldwide.*

All this is part of the culture of sound management. The conceptual framework underpinning the described approach is crucial in implementing a chosen strategy. Its aim is to permit effective integration of corporate goals with ways and means, presently available and under development. This is essentially the interface of strategy and technology.

In sharp contrast to this wider but focused approach by Bezos has been the one followed by Robert E. Allen, AT&T's CEO in the 1990s. He launched half-dozen poorly coordinated local-phone strategies, from simply reselling services bought from incumbents to a new system for "fixed wireless" (primarily a residential service)—all at high cost and low returns.

In 1996–1997, under Allen, AT&T dramatically increased capital spending from an annual budget of $6.6 billion slated for building local networks. Still none of the approaches, save the SBC merger, appeared to have

---

* D. N. Chorafas, *Cloud Computing Strategies*, Auerbach/CRC, New York, 2010.

the potential to get AT&T into the $100 billion local U.S. telephone market in a big way. Said Jeffrey Kagan of Kagan Telecom Associates, "They're not anywhere near where they wanted to be or where analysts expected them to be."*

Time and again sources close to AT&T commented that its executives were pursuing "this" or "that" other acquisition, as well as marketing deals with small companies that competed with the Bells, without having a strategic plan. The short-term goal was that of linking business customers to long-distance networks.

This left Wall Street totally unconvinced about AT&T's policies and moves in the mid-1990s. A frequently heard comment by financial analysts was that such deals will never make AT&T a local power in U.S. telephony. It's like stringing tin cans together and tying it with a knot, said one of the analysts.

A cause and effect analysis would reveal that at the origin of this stringing of tin cans has been the fact that prior to the dismantling of AT&T, by decision of Judge Green, the company was a huge bureaucracy. Its managers were highly paid bureaucrats running a secure industry and found themselves like fish on a dry floor when they had to confront market forces.

Deutsche Telekom, France Telecom, and Holland's KPN are other examples of former public utilities whose bureaucrats went berserk when they were let loose. KPN was saved from bankruptcy through taxpayers' money; by most significantly increasing their leverage, the German and French telecoms landed in a sea of troubles.

- These strategic mistakes are current currency among bureaucrats.
- Entrepreneurs with the salt of the earth also make mistakes, but they correct them much faster and they know how to come up from under.

An argument often heard is that AT&T, KPN, Deutsche Telekom, and France Telecom were obliged to compete by delivering services others, too, could deliver. This is holding no water. Amazon.com is just one of several companies, which won by delivering services that others can also deliver, but it reached leadership because it had strategic direction, low cost, fast response, and a more efficient solution that provided it with market leadership.

---

* *BusinessWeek*, July 17, 1997.

## 1.6 DEVISING A STRATEGY FOR GROWTH

Like Silicon Valley companies, the formerly lethargic industry of auto manufacturers is now on the go. The awakening has come through globalization, which made Detroit's massive, slow-moving "motor town" policies the best recipe for losing one's market and getting out of business.

- The motor vehicle industry has become one in need of new ideas, methods, and business models.
- The key is more efficient, sophisticated, and eco-friendly designs and materials.

Only by working with strategic partners and sharing innovative engineering concepts will motor vehicle companies find a valid solution. Well-managed firms know that design engineering is at the heart of the auto recovery, and that an equally potent factor is the ability to reach the global market. Hence the thirst not only for new technologies but also for size.

In the post–World War II years Fiat was *the* motor vehicle manufacturer of Italy, eventually absorbing its competitors Lancia and Alfa-Romeo. But this did not make the company a global player, even if on and off it established production facilities (and local labels) in France and Spain.

In the 1990s Fiat tried to get around this problem with various alliances. The better known is a tie-up with GM that lasted for 5 years, from 2000 to 2005. It ended in 2005, and Sergio Marchionne, Fiat's new CEO, extracted $2 billion from GM to extinguish a put option that would have forced GM to buy the Italian company, which was then teetering at the edge of the abyss.

Fiat was in a fight for its life. It knew that it did not have the resources and global market appeal to be one of mass producers like Toyota, Ford, GM, or Volkswagen. Nor had it the breadth of time for an alliance of French–Japanese style—like Renault–Nissan. For Fiat, 2005 was a strategic inflection point. Small size can be deadly in an industry with global production capacity nearing 100 million vehicles per year, far in excess of demand.

- In good times the global market absorbed about 60 million.
- In bad times during the economic crisis from 2007 to early 2009 (when government-sponsored promotion programs got into effect) the global market shrank to 45 million.

Moreover, not only Fiat's own experience but also that of other car makers had led Marchionne to believe that alliances are all very well, but they react too slowly, involve too much horse trading, and require too many compromises to be rewarding in the longer term. Cars have become an industry where speed in decision and execution is at a premium.

During 2008, Sergio Marchionne had been developing an ambitious strategy. First it was constructed around Chrysler, then expanded to include Opel. Badly mismanaged by Daimler during its decade of ownership, and too dependent on pickup trucks and aging car models, Chrysler was in no condition to survive if it stood alone. Its rather poor management sought help from other carmakers, including Fiat and Nissan.*

Marchionne devised a plan to win Chrysler the federal loans it needed to stay alive, while putting Fiat's arms around the U.S. car maker. In exchange for an equity stake of about 35 percent, Fiat proposed to make available to Chrysler its small- and medium-sized platforms and advanced, fuel-efficient powertrains. In exchange, Chrysler would give Fiat some of the scale it was seeking for its platforms, as well as

- joint purchasing of parts
- expertise in producing large cars, and
- most importantly, a U.S. distribution network

The scheme won the backing of the Treasury's car industry task force. Although the plan established through negotiations called for Fiat to initially get only a stake of 20 percent, this was to rise to 35 percent after fulfilling criteria set by the Treasury. But the Italian company would have to repay all Chrysler's federal loans before taking majority control—a heavy duty indeed.†

The American Treasury accepted that Fiat should take over the responsibility of managing Chrysler. Fiat's management was also to be responsible for integrating the two organizations—while restructuring and revitalizing Chrysler, the way Marchionne had done at Fiat back in 2004 when appointed its chief executive.

* Nissan would have made a small car for Chrysler and Chrysler would have built a pickup for Nissan.

† It is interesting to note that in 2008 Fiat and Chrysler were neck to neck in auto production, each with slightly over 2 million cars per year but way behind other competitors—from Toyota to Renault.

Opel was Fiat's next challenge. In early 2009 a badly wounded GM sought to retrench, and this seemed to include letting lose Opel, its huge European subsidiary. To save jobs for cars "Made in Germany" the German government contemplated stamping up $5 billion in taxpayer money in bridging loans. This was supposed to keep Opel's factories open.

Marchionne saw in Opel an opportunity for Fiat to gain size, as well as part of the German market. As suitor for Opel, he promised to keep open its three main assembly plants in Germany, although factories in Belgium and Britain might have been closed. Over time he intended to reduce combined capacity by 22 percent, but he said he will do so by slimming factories rather than closing them—because, as he put it, this is the preferred way in Europe.

Cost savings were part of the strategy. By combining Opel and Fiat it might have been possible to save at least €1 billion ($1.5 billion) a year. The two companies also complemented one another. Opel lacked an A platform, which Fiat had, and they shared a B platform for the Corsa and the Grande Punto. Fiat, moreover, could use Opel's new C and D platforms.

All that sounded very interesting, but it was far from sure that Fiat pulled it off, because GM was unsure about whether or not it wanted to keep only a minority interest in Opel. In addition, there were other suitors for Opel. Politics, too, played a role. Germany was run by a coalition government:

- The Christian Democrats favored an Italian solution for Opel.
- But the Social Democrats preferred the offer of Magna, a Canadian–Austrian autoparts producer, in alliance with Russia's Sberbank.

Financial analysts were not particularly enthused by the attempt to meld Fiat, Opel, and Chrysler because of the formidable problems it presented.* By contrast, this seemed like a perfect opportunity for its promoters who deployed their connections in politics and business to see it through.

Eventually, Magna seemed to carry the Opel prize. In fact, it was announced as *fait accompli*. But contrary to the information that was widely distributed, this was not a done deal because GM had not yet given its seal of approval. A palace revolution at GM's board changed the whole scenery. Opel remained a fully-owned GM subsidiary, after all.

* Critics doubted that the merging of three different cultures and equal number of technologies would work.

As the Fiat, Chrysler, Opel, and Magna twists demonstrate, while strategic plans evolve, different deals among automakers are done and undone—while production capacity is far in excess of what can be justified by the market's needs. In addition, not everything is connected to auto production. Among the better managed auto companies attention now shifts toward the dealer network—which has been neglected for too long.

In a tie-up with Berger, a consultancy, and the Politecnico di Milano School of Management, Fiat launched an intensive training program for its network of distributors. It started in Italy, expanding to Germany and

**TABLE 1.1**

Strategic Perspectives and "This Year's" Thrust Established by the Board of a Major Financial Institution

| | Overall | "This Year's" Thrust |
|---|---|---|
| 1 | Devise a sound strategy to improve returns, and implement it successfully | Function like a merchant bank |
| 2 | Develop uptier corporate relationships | Improve personal contact with senior management of correspondent banks and major clients |
| 3 | Service large domestic and multinational corporations at home and on global basis | Develop noncredit services as lending becomes more competitive |
| 4 | Expand global personal banking activities | Assure strong fee income growth, at slower expense growth |
| 5 | Build greater expertise in corporate, fiduciary, money/capital, and investment banking, to develop full-fledged bilateral trades | Expand revenue from operational products through focused sales and trading |
| 6 | Profit from trading and funding skills in volatile markets | Selectively expand relationships with foreign multinationals and financial institutions |
| 7 | Improve customer profitability by increasing contact and reducing service costs | Use leading-edge technology to support sophisticated services while cutting costs |
| 8 | Enhance the ability to attract and retain some of the best financial minds in the world | Retain good earnings visibility based on strong net interest margins, and reward performance |
| 9 | Upgrade the risk management practices, and make all managers and professionals risk conscious | Develop a fully integrated risk management system, supported by knowledge engineering with risk positions updated in real time |

Britain. These management courses (which also target Fiat managers who work with the dealers) are spread over eight two-day sessions spread over eight months. They cover

- managerial and entrepreneurial skills, and
- innovation and organizational principles and practices.

The aim is to increase the participants' ability not only in selling cars but also in managing and controlling their enterprises. The latter is, unquestionably a domain in which all marketing and sales people should undergo intensive training. The best strategy for survival is to develop and maintain human resources.

All industrial sectors must not only be keen to steadily upgrade the skills of their people entrusted with sales and customer handholding, but the financial industry also is confronted with the same need. Table 1.1 presents a snapshot of the nine top strategic perspectives established by the board of a bank to which I was consultant. Nearly half are particularly focused on human capital. (See also Chapters 2 and 6.)

# 2

# *Management*

## 2.1 THE FIRST SYLLABLE OF "MANAGEMENT" IS "MAN"

The first syllable of *management* is MAN.* In addition, like the term Fortran derives itself from FOR-TRANSIT, the term management can also be read

MANAGE – MEN – TACTFULLY

Tactfully means many things, one of them treating people equally. "Do you like Vince Lombardi (the well-known football coach)," asked a reporter to one of the players. "Yes," was the answer. "But Lombardi treats you like a dog," insisted the reporter. "Yes," said the player, "but Lombardi treats everybody like a dog."

"Like a dog" is of course a figure of speech, but it is no less true that laxity is *mismanagement.* For eons, laxity and management are at each other's throat. An analysis of the history of business failures demonstrates that adverse events happen when management is too weak, indecisive, or lax. In the aftermath

- the firm fails to serve its market(s) in the best possible way
- competitors encroach in and/or take the lead decreasing the firm's business, and
- customers, particularly big customers, escape the firm's hold when its management becomes big headed as they hate that attitude

Being lax, weak, or big-headed are sure signs of decay both of the company and of the people working for it. Contrary to whatever the theorists like to think and say, management is not a "be good" and "be

* I have toyed with the idea of writing "man or woman" but in this context it would not have made sense. Interpret it, however, *as if* it were written that way. Similarly, *he* means "he" or "she."

loved" business—neither is it a science, for that matter. First-class managers are those able to take tough decisions, whatever these decisions *need* to be.

Tough decisions made by MAN are necessary because adversities, dilemmas, and difficult trade-offs crop up time and again. "Tough" does not mean "inhuman." Companies are human organizations rather than just sources for economic data, as the late Peter Drucker has explained on several occasions.

Drucker insisted that all human organizations, whether in business, in government, or in voluntary sectors need clear objectives as well as realistic, hard measurements to keep them efficient. He also pressed the point that managers have to make tough decisions without a hint of embarrassment—because, after all, that's their job.

No matter what might be the background reason, the avoidance of clear and unambiguous decisions leads to disastrous results. They don't allow organizations to move ahead with their time, into new spheres of activity for fear of being confronted by

- resistance to new ideas
- little known methods or tools, and
- unconventional modes of life

Resistance to change makes more expensive an adaptation when this becomes due. It is often said that generals fight a new war with the weapons, methods, and ideas of the last war—and there is plenty of truth in it.

This is particularly true in the military, the government, and corporate life when the top brass are second raters who live in the past and not in the future. By contrast, leaders in the art of war and of any other activity know that they have to

- feed their strengths and opportunities, and
- starve their weaknesses wherever they may lie

Both require tough decisions, typically taken by *first raters*. In the background of every first rater is MAN. In his short but powerful book *Essays on Science*, Albert Einstein wrote that men of action in any sphere of human activity will do. They will always show up in positions of leadership, and whether they become scientists, businessmen, traders, high priests, or military geniuses often depends on circumstances.

> The state of mind which enables a man to do work of this kind is akin to that of the religious worshipper or the lover; the daily effort comes from no deliberate intention or program, but straight from the heart.*

Renowned military leaders have been first and foremost great managers of human resources and of all other assets. This also describes hand-in-glove what lies behind the achievements of business tycoons. Therefore, a great deal can be learned in management practice from the life and principles of conquerors.

How they achieved what they did, should not be obscured by the fact that history has condemned most conquerors to miserable, untimely deaths. Alexander the Great died under mysterious circumstances, probably poisoned by his generals. Caesar was murdered in the Roman Senate by his opponents, and by Brutus one of his former lieutenants. Napoleon faced death as a solitary prisoner in the island of Santa Helena, allegedly poisoned by one of his generals in British service.

Genghis Khan, by contrast, the way Jack Weatherford puts it, passed away surrounded by a loving family, faithful friends, and loyal soldiers ready to risk their lives at his command.† In many countries, and particularly so in the Muslim world, Genghis Khan is considered as a terrible person. This, however, is far from being a universal opinion—or the opinion the reader makes by reading his biography.

Not only do the Mongols have him in high esteem, but in China as well he is revered as the founder of a dynasty.‡ In management theory, Genghis Khan ranks high because he recognized that warfare was not a sporting contest or a match among rivals fighting in a ring. He expressed the management concepts at the top of his mind in five *principles of leadership*:

1. The first is *Self-Control*—particularly the mastery of pride:

   > If you cannot master your pride, you cannot lead. Mastering your pride is more difficult than fighting with a lion.

---

* Albert Einstein, *Essays on Science*, Philosophical Library, New York, 1934; English translation of Einstein's *Mein Weltbild*, Querido Verlag, Amsterdam, 1933.

† Jack Weatherford, *Genghis Khan*, Three Rivers Press, New York, 2004.

‡ It is interesting to note that the Mongolian term *Hun* often used in the West as synonymous to aggressiveness and an uncivilized behavior, means *human being*.

2. The second is *Vision, Goals, and a Plan:*

   Without vision of a goal one cannot manage his life, much less the life of others.

3. The third is to *Conquer the Hearts of People:*

   You can conquer a nation only by conquering the hearts of its people.

4. The fourth is *Simple Style of Living:*

   If you are slave to luxury, you will be no better than any slave.

5. The fifth is to *Demonstrate Thoughts and Opinions through Actions*—not through words:

   Don't talk too much. Only say what needs to be said. Game slaughtered through words cannot be skinned.

In the background of these principles lies the concept that leadership and management are a total commitment, not a part-time business. Victory comes to the one able to impose the rules that he has made. This, says Weatherford, Genghis Khan derived not from some sort of enlightenment, but from a persistent cycle of *constant revision* driven by his uniquely disciplined mind and focused will.

## 2.2 LEADERSHIP

A man, an organization, and the whole of society get paid for their strengths—not for their weaknesses.* What's *my* strength? This is a crucial question every person should ask himself or herself. To a substantial extent, strength is measured by performance, and performance is the result of the steady drive in applying one's know-how and capabilities to the work being done.

Leadership is a personal characteristic. People usually associate leadership with success, but the two terms are not synonymous. Success is a matter of concentration on areas of strength and of quality of deliverables. "Only the strong can be productive and only the productive can be free," said John Foster Dulles, the Secretary of State in the Eisenhower administration.

* There is an exception to this statement. Bought souls are paid for their weakness of being bought and sold—in a literary sense.

- In principle, the future is determined by the goals we establish for tomorrow and by our ability to meet them.
- Leadership's role is to promote the attitude of being ahead of the curve. People and organizations successful in yesterday's realities, tend to be defensive and fall behind the curve.

Two types of leadership characterize every enterprise, from business and industry to politics. The one is the so-called *charismatic*, full of promises, and lots of glitz and glamour combined with the ability to sustain them over time. This type of leadership finds its roots in the personal strategy of going around shouting the odds. Luck helps, but this is only one factor—and it is perishable.

Surely it is not just "luck" which saw to it that at the dawn of the twentieth century three men, Henry Ford, John Rockefeller, and Andrew Carnegie, controlled among themselves the whole industrial might of America. Neither is it "destiny" that turned them into some charismatic leaders of industry.

The other type is leadership based on *fundamentals*. People and companies who follow it put in place firm policies that are well founded and to which they stick. Steve Jobs and his company, Apple Computer, are examples. They lead by example, which says "This is me, this is who we are, and this is what we stand for. We can show you what we are doing, and what we are doing is effective."

Amazon.com's Jeff Bezos provides another excellent example on how to manage a company based on fundamentals. From 2006 to 2009, in a period largely characterized by a deep economic and financial crisis, *his firm* had a stellar performance:

- Business grew by 114 percent.
- Profits grew by 374 percent (much faster than business).
- Personnel alone grew by 72 percent (only a fraction of business growth).

To a most substantial extent, this has been the policy followed by all of the promoters of industry. In the early stages of the development of the modern corporation there was always an innovative and dominant personality whose strategy and acts have been based on fundamentals. Examples are as follows:

- Carnegie at Carnegie Steel
- Morgan at Morgan Bank

- Rockefeller at Standard Oil
- Ford at Ford Motor Company
- Sloan at General Motors
- Firestone at Firestone Tires
- Watson at IBM
- DuPont at DuPont
- Sewell Avery at Montgomery Ward*
- Rosenwald at Sears
- Giannini at Bank of America
- Juan Trippe at PanAm

An originating name can be identified with nearly every one of the Fortune 500 corporations in manufacturing, merchandising, transportation, or finance. At a second stage, however, many prosperous organizations were taken over by mediocrity; the second raters dominated and when that happens the company goes into decline.

Among more recent cases of people and companies based on fundamentals, Bob Metcalfe is the example of an industrial and industrious person whose leadership has left a footprint. "Some call it luck," says Vint Cerf, one of Arpanet's (and therefore the Internet's) developers. "But Bob has an ability to detect and take opportunities. And he is willing to take risk."† Metcalfe created a new networking technology, the Ethernet, that revolutionized computing. His assets have been his ability to

- observe
- synthesize, and
- improve things that others would have looked at without any motivation

Local area networks, like the Ethernet, brought forward the client-server computing model, which significantly increased cost effectiveness and demolished the saga of the mainframes (even if second raters at many companies continue to believe that it is nice to live in the Middle ages of computing). There were as well missed opportunities.

Bill Krause, who worked at 3Com, Metcalfe's outfit, in its early days refers to a "100 billion dollor mistake" that he and the company's boss

* Though late in his career, in the post–World War II years, he aged in his thinking, fell behind and lost to Sears.

† *The Economist Technology Quarterly*, December 12, 2008.

made in overlooking the gold mine of linking entire networks together—rather than just computers. In Krause's words "A little company called Cisco" saw that bit of the future and became the main supplier for "networks of networks."*

We are all bound to now and then make mistakes. The alternative is remaining inactive, and it is unacceptable. Speaking from personal experience, the best entrepreneurs I have met are those who learned quickly from their mistakes and did not repeat them. They took risks, but they knew and understood the risks they assumed, putting limits to them and watched them carefully, correcting the deviations when and where they happened.

This strategy is inner-directed. It abides by Miyamoto Musashi's techniques, which the famous Samurai characterized as speaking to the soul of the individual. The approach followed by Musashi has been study, meditation, and mental preparation. "My way of strategy is the sure method to win when fighting for your life, one man against 5 or 10," Musashi says in *A Book of Five Rings*.†

Musashi talks of a *culture*, not just of an individual behavior. This culture rests on ethics and on the ability to deliver by overcoming adversity. "Difficulties" is one excuse by second raters that economic history never accepts. Development and survival depends on vigorous interaction with the environment, which reflects

- staying power
- superior know-how
- willingness to adapt
- ability to change, and
- endurance in the face of adversity

These are the forces that help men become masters of their fates. Unfortunately, they are not the themes taught at schools and universities, neither can they be found in cookbooks about management. That's what teaches the Mahendra‡ hypothesis on the weaknesses of business schools and in the background they provide to their students.

---

* Idem.

† Miyamoto Musashi, *A Book of Five Rings*, Overlook Press, New York, 1982.

‡ Mahendra is an Indian guru and rather successful financial forecaster. See also the reference to Bernheim in Section 2.3.

The tools, models, and theories business schools practice—be it economic, financial, quantitative, or technical—are never more than just small parts of a gigantic bowl. What is missing is the forward looking management notion, which is found at the base of leadership. Business schools' teachings and associated practices, Mahendra says,

- are not looking beyond conventional methods
- Do not challenge the 'obvious,' and
- Fail in prodding the fundamentals largely because of egos and pride

Big egos and overwhelming pride are dysfunctional (see also the reference to Genghis Khan's management principles in Section 2.1). In terms of business teaching, over the years they led to the fact that conventional investment strategies remained focused on quantitative economic data and associated news items. By contrast, both the economy and the market's fundamentals include a wealth of factors not being accounted for.

- Markets are trading with huge uncertainty and volatility.
- While traders, analysts, and investors try every time to find bottom in stocks and commodities.

Countless financial professionals have not been coming out of school with any advantageous model that can guide them and their clients well over the long term. After studying all the theories and models, they still don't have anything unique to offer, at least nothing that comes close to the fees they charge and risks they take. What is missing is leadership.

## 2.3 THE MANAGER OF THE TWENTY-FIRST CENTURY

The company of the twenty-first century will be rich in professionals, and proportionally thin in managers, because the role of management is changing so much. In a way, this new entity will resemble an orchestra with 200 or 300 professionals and one conductor, according to Peter Drucker.

An orchestra does not have sub-conductors, the way industrial and financial organizations are built today with plenty of middle management layers. But an orchestra has a first violin, a top professional who distinguishes himself by being a *virtuoso.* As for the conductor himself, he is a *doer* characterized by

- conceptual capabilities, and
- directional skills associated to them

To appreciate the deeper meaning of conceptual capabilities and their aftereffects, presidents, board members, managers, professionals, and employees will be well advised to study the life of the great *doers* of mankind. One of them was Henry J. Kaiser, the American industrialist who in World War II built cargo ships in five working days a piece, when the best shipyards at that time required 210 days.

Kaiser's motto was "You find your key men by piling work on them."* This search for talent is exactly what a modern company should expect from its CEO—its orchestra conductor. At a construction site, following a violent storm, the earth-moving machinery was buried in a sea of mud. "Whatever are we going to do?" a workman said glumly to Kaiser, "Just look at this mud." "What mud?" Kaiser asked, "I see only that big sun shining down. It will soon turn that mud into solid ground." In a nutshell, that's why Henry Kaiser accomplished so many seemingly impossible things—he saw opportunities in every difficulty, rather than difficulties in every opportunity.

Like the modern manager should do, Kaiser had a genius for getting the most of other peoples' talents, and then giving them credit. He was not egoistic. He shared the credit, even when he pioneered the vertical and horizontal integration of allied industries. Building dams, he manufactured cement. Building ships, he made his own steel. Most importantly, he surrounded himself with talented young executives, thrusting upon them responsibilities beyond their years. He did so because he

- paid attention to human creativity in labor and management, and
- believed that the impressions a young man gets when he first starts in life have much to do with the shaping of his life's success

By any standard, Kaiser's business philosophy was unique. He felt that it is more important building people than building dams or plants. His approach was unconventional in the 1930s and 1940s but may well become

* Albert P. Heiner, *Henry J. Kaiser, A Western Colossus*, Halo Books, San Francisco, CA, 1953.

mainstream in this century. "I have never lived according to conventions," he said in an April 8, 1951 interview to the *Oakland Tribune*, "I have defied the conventions all my life. I cannot stand not to be busy, not to work."*

In terms of deliverables, Henry Kaiser's credo was that a little ingenuity and a lot of hard work could overcome almost any mechanical problem. He himself constantly tinkered with every piece of equipment he worked with, trying to find a way to fulfill the job

- faster and
- more efficiently

Doing so, he capitalized on his conviction that the faster a project was, completed the lower the cost would be,† even if at times extra expenditures were needed to bring about the step-up in speed of execution. His keywords were *work* and *economy*, and to honor them he kept on digging till he understood the problem.

Though Henry J. Kaiser's high water marks were the 1930s and 1940s, more than two-thirds of a century ago, the management principles he established may well be forerunners of those characterizing this century. *Work* and *economy* are the dual pillars on which a person, a company, and a nation can stand in times of turbulence. At the roots of Kaiser's philosophy is found the belief that

- to a substantial extent trouble is predictable, and
- capable managers face tomorrow's challenge sooner rather than later

This calls for a steady process of reevaluation and upgrade of products, plants, offices, and above all human resources. Also for critical examination of changes taking place in the client base. For instance, the more a country industrializes, the more its customers become other industrialized countries—and a company's clients dispose more negotiating skills than ordinary.

This brings into the picture the strategy of seeking partners and forging alliances. In its effort to reduce the cost of developing an innovative new aircraft, Boeing recruited risk-sharing partners who became largely

* See also Section 2.5 on hard work, and Edison's dictum.

† Costs matter, the more so in a globalized economy. On the importance of cost control, see Chapter 15.

responsible for designing whole sections of the plane. The concept of risk-sharing was brilliant, but it also created one of the most extended and complex supply chains in industrial history.

Boeing's mistake has been that it failed to supervise its partners' work adequately. Critics say that the deal delayed the Dreamliner's production as well as ended in spending more to put things right, than it ever would have saved. Brilliant new ideas are often killed by management's defective oversight—which is indeed one of the most deadly weaknesses of twenty-first-century management.

In this, there is no better advice than that given by Alfred P. Sloan: *React Quickly.* As chairman of GM, Sloan gave an excellent example on the need to be ready and take no delays, when he steered GM in a way to avoid the aftermath of the 1929–1932 Depression suffered by other companies: "No more than anyone else did we see the depression coming..." Sloan said. "We had simply learned how to react quickly. This was perhaps the greatest payoff of our system of financial and operating controls."*

In the deep economic and banking crisis of 2007–2010, JP Morgan Chase managed to avoid big losses largely thanks to culture set by its CEO Jamie Dimon. PowerPoint presentations were discouraged, informal discussions of what is wrong, or could go wrong, encouraged—and there was a hard-headed approach to the allocation of capital.

Risk management culture may look like an elusive concept, but it matters. "Whatever causes the next crisis, it will be different, so you need something that can deal with the unexpected. That's culture," says Colm Kelleher of Morgan Stanley† One necessary ingredient is a tradition of asking and repeating questions until a clear answer emerges.

Asking pointed questions and expecting clear answers is a characteristic of managers who have the salt of the earth. In his excellent book *Adventures of a Bystander*, Peter Drucker mentions the story of Henry Bernheim, the man who started from nothing and, at the end of the nineteenth century, made one of the most important merchandising chains in the United States. Bernheim had sent his son to the then brand new Harvard Business School to learn the secrets of management.

"But father," said the son when, on his return from higher education, he was given a chance to look at the department store's books, "You don't even know how much profit you are making." "Come along my boy,"

* Alfred P. Sloan, *My Years With General Motors*, Sidgwick and Jackson, London, 1965.

† *The Economist*, February 13, 2010.

answered Henry, and brought him on an inspection tour from the department store's top floor to a sub-sub-basement that was cut out of bedrock. There, at rock's edge, lay a bolt of cloth.

True enough, contrary to what business schools teach, there were no statistics around, no balance sheets, and no income statements at Bernheim's. But there was that bolt of cloth. "This is what I started with," said the father. "All the rest it's the profit."* (See also in Section 2.2 the Mahendra hypothesis on business education.)

## 2.4 MANAGEMENT DECISIONS

In a way reminiscent of Henry J. Kaiser's principles and policies, today successful chief executives are driving projects by keeping their people on the run. They care for, but they don't worry about tomorrow. The best guide to this is the advice given by the father of Admiral Chester W. Nimitz to his son: "The sea—like life itself—is a stern taskmaster. The best way to get along with either is to learn all you can. Then you do your best and don't worry—especially about things over which you have no control."†

This applies hand-in-glove to the *decisions* to be made. These are acts preceded by deliberation, calculation, and thought. For 16 years I was personal consultant to one of the renowned leaders of industry and from him I learned the importance of *focus by management*, prior to reaching a decision. Focus is first and foremost what Carlo Pesenti expected from the presidents of his industrial companies and financial institutions. To his mind

- it is more important to focus on an issue
- than to be "right" about the outcome of a decision

Focusing on an issue is the hallmark of a sound managerial attitude, and very often it conditions the nature—of the decision itself. Take the case of a start-up in technology confronted with the four quarter spaces shown in Figure 2.1.

* Peter F. Drucker, *Adventures of a Bystander*, Heinemann, London, 1978.

† John Wooley and Gerhard Peters, "Gerald R. Ford. Remarks at the U.S.S. Nimitz Commissioning Ceremony in Norfolk, Virginia," http://www.nimitz-museum.org/nimitzbio.htm

*Strategic advantages*

| *Strategic market target* | Unique product | Product leadership |
|---|---|---|
| Wide market | Differentiate | Be compatible |
| Niche | Focus | Transit to wider market |

**FIGURE 2.1**
No choice is self-evident. A priori management decisions must be focussed not abstract.

- If it goes for a unique product and a wide market, it must bet on being different, not just vaguely "better."
- But if the strategic choice is that of a niche, it must concentrate on competitive advantages, which will allow it to write its ticket.
- Alternatively, it may bet for low cost leadership, which in a wide market would mean compatibility (Chapter 13), making a niche market choice only be a temporary solution.

The principle of *focus* applies to all problems, big and small, because no problem is "too big" to run away from. Failure to focus on the subject being decided and on the contemplated decision's aftereffects is destroying the firm in the longer run. Focus, however, requires thinking, and as Edith Hamilton has it "People hate being made to think, above all upon fundamental problems."*

- *Thinking* is, so to speak, movement without movement.
- We contemplate then simulate a move, try to guess what it involves and, if possible, project its aftereffect.

This may sound simple, but it is complex and demanding. It also requires lots of training. One hundred years of evolution have been devoted to shaping up the process of thinking, particularly the network that controls the brain functions and whose exact pattern is still unknown (including matters regarding behavior).

* Edith Hamilton, *The Greek Way*, W.W. Norton, New York, 1930.

Decisions without focus and without thinking are the worst pastime possible. Therefore a great deal of effort is today expanded in reverse engineering the thinking process along the hypothesis that

- the nature of learning and of all sorts of human activity, including aggression, are keys to understanding human behavior, and
- starting from such understanding, it may be possible to unlock some of the basic aspects of thinking and of its impact on decisions.

Far from being an academic exercise, this is a most practical endeavor because the importance of thinking is as great in management decisions as in research and any other creative activity. In his Cambridge lecture on August 31, 1837, Ralph Emerson said, "*If* one is a true scientist, *then* he is one who THINKS."

Niels Bohr, the nuclear physicist, was teasing his peers and his students by telling them "You are not thinking, you are only being logical." Great men in history have always appreciated that thinking means challenging the "obvious," therefore analyzing, simulating, and experimenting. Because it is based on thinking, experimentation is the mother of all sciences.

Management, of course, is not a science, no matter what long-haired people say. More than anything else it is an art, but analysis that should precede a decision is both scientific and integral part of the managerial function. The need for analysis is present whether a decision has to do with a business opportunity, competitors actions, communications received from a government authority, important cases referred to by subordinates, or is connected to the executive's own initiative. In all cases, decisions must be

- timely
- pertinent to the subject at hand
- executable; that is realistic, and
- within the authority of the person making them

Leaders of industry say that *timing* is everything, and they are right. A decision followed by an action taken early, may well fall short at a later day because what it stipulates would not be enough. The $150 billion in loans (€110 billion) provided by Euroland's governments and the IMF to Greece would have been much more than enough in November 2009 and they were short by half when they were offered in May 2010.

Many managers try to justify their inaction with the excuse that they are "waiting for more information." Too much information, however, may be as bad as too little. Herbert Simon, an economist, said "What information consumes is rather obvious: it consumes the attention of its recipients…a wealth of information creates a poverty of attention."

The right kind of timely and accurate information is however necessary because management decisions deal with the allocation of limited (finite) resources in men, money, and means to carry on operations. They are made in the face of uncertainty, and a recurrent theme in decisions is to preserve these resources or, even better, increase them.

This is one of the basic reasons why the ability to conduct experimentation on alternative possibilities and outcomes of the *what if* question is getting appreciated by management. Decision theory applies statistical analysis to problems characterized by uncertainty. The pivot point is assignment of probabilities of occurrence to events, along with estimates of payoffs or losses that would result if the event occurs. Analysis then centers on the search for the best possible payoff (or least loss).

The importance of decision theory and of analytics at large lies in bringing home the message that the process of management is not just one of assigning authority and responsibility, but also (and greatly) involves studying, altering, and exploring all that leads to a choice. A prerequisite to do so is identifying areas requiring attention, as well as their characteristics and pitfalls.

The importance of establishing and maintaining a sound process of management decisions is not always appreciated. Yet, experience shows that it can be instrumental in changing the thinking of an organization. An orderly approach is characterized by a step-by-step procedure:

- State the objective.
- Examine alternative paths.
- Work out how to get there.
- Qualify and quantify the benefits.
- Outline adverse reactions, limits, and risks.
- Cost out everything.
- Systematically monitor progress against the plan.
- Reward effectiveness and efficiency in execution.

Effectiveness and efficiency are not the same thing, as is sometimes suggested. Peter Drucker has properly labeled them, as shown in Table 2.1,

**TABLE 2.1**

Efficiency and Effectiveness

| Efficiency | Effectiveness |
|---|---|
| Doing a given job at least possible<br>• Cost<br>• Time | Deliverables meeting goals in a most able manner |

giving each concept its due. He also said that "Effectiveness is a habit," and he is right. Management decisions must be effective, and they should be executed in an efficient way.

At the early part of his career at Ford Motor Company, Robert McNamara provided a good example on how to reach effective and factual decisions: through value analysis. To increase the cost/effectiveness of the Ford model all the components of each new Chevy, made by GM, would be laid out for competitive evaluation.*

Most evidently, behind all goals, decision making, analytics, evaluations, and cost-benefit studies stand ordinary human beings, and many (some people say all of them) behave unpredictably. This is a basic reason why management is an art—though decisions can greatly benefit from salt-of-the-earth analytics.

## 2.5 HARD WORK

Thomas Edison is credited for having said "Success is 2 percent inspiration and 98 percent perspiration." Nothing can be achieved without the work ethic, or by means of trying to avoid hard work. This is what every father should teach his son(s) and daughter(s).

When Robert Taylor asked Louis B. Mayer, the boss of MGM, for a raise, L.B. advised the young man to work hard, respect his elders, and in due time he would get everything he deserved. To the question asked by his colleagues "Did you get your raise?" the tearful Taylor is said to have answered "No, but I found a father."†

* At the Air Force Office of Statistical Control, where he worked in war years, McNamara counted the firebombing sorties made by the B-29s, at what height, with what percentage hits on target: 8 percent of Yokohama, 51 percent of Tokyo.

† *Time*, December 7, 1998.

Andrew Carnegie started in life by working 6 days a week at $1 per day, as a young teenager, studying in the seventh grade with books borrowed from the local library. His first breakthrough came at an early age when he landed in a job as secretary and telegrapher to Tom Scott, a powerful stakeholder of the Pennsylvania Railroad. At age 23, Carnegie headed Penny's Pittsburgh division and began to collect a small fortune from outside investments ranging from oil to iron bridges.

Carnegie's pursuit of industry was focused and therefore contrarian to today's rush to diversification. "Put all your eggs into one basket," he once advised, "and then watch that basket." For him the basket was steel. As an investor and manager he was

- fiercely competitive,
- able to make tough decisions, and
- bent on innovation and efficiency

Like all princes of industry, Carnegie worked 15 hours per day, 7 days a week. His decisions were fundamental. He would scrap a relatively new plant to erect a more modern one because he had learned from experience that, by 1990, innovation permitted him to reduce the price of rails—the product that initially drove the steel industry—to $17 a ton from $160 a ton in 1875.

After selling Carnegie Steel (now U.S. Steel) to J.P. Morgan in 1910 for $480 million (a huge amount at that time) Carnegie devoted himself to an active form of philanthropy. In remembrance of the fact that he got his learning in public libraries, he created 2800 free libraries worldwide, endowed the Carnegie Corporation, made the Carnegie Endowment for Peace, and a set of a dozen of other major philanthropic organizations.

All that with the capital he had earned through his work. Over the years he was in charge of his individual empire, Carnegie appreciated that the man who commands the enterprise is not a mere technician. He is the animator, promoter, planner, and organizer of the enterprise—the person who values above all the character qualities of

- hard work,
- creative imagination, and
- unbiased judgment

These are not rights. They are responsibilities. It is the CEO's responsibility to assure his company's profitability, customer satisfaction, and internal control. All three combine to give the firm a unique competitive edge. It has been Lee Iacocca's* business philosophy that all business operations can be reduced to three words:

- People
- Products
- Profits

Because people are the basic ingredient, in planning and in selling, Iacocca has suggested to always think in terms of the other person's interest. But interest and credibility are not the same thing. Credibility is something one can only earn over time—and if one has not earned it, one cannot use it. Building up credibility requires a lot of hard work and of deliverables.

In a similar manner, innovation requires a lot of imagination. In Iacocca's opinion successful products are innovative, easy to manufacture (which he sees as the key to *quality*) and low cost in direct labor and direct materials (Chapter 15). Quality and productivity, he advises, are two sides of the same coin and both involve the CEO's personal responsibility.

Responsibility is the *alter ego* of managerial position, and it is indivisible from authority. Once authority has been accepted, responsibility sets in. As John D. Rockefeller, Jr. once suggested, I believe that every right implies a responsibility; every opportunity, an obligation; every possession, a duty."

Significantly, in Carnegie's, Iacocca's, and Rockefeller's statements is missing the word "rights," which distorts the sense of responsibilities—because responsibilities come way ahead of rights. Instead, the emphasis is on the hallmarks of competitiveness: a well-conceived strategy, hard work, innovation, steady product development, focused marketing, and emphasis on organizational ingenuity. A lean structure is needed in order to

- improve the process of decisions,
- get closer to customers,

* Formerly CEO of Ford, then of Chrysler, and prior to that the man who projected the characteristics of a popular sports car which led to designing Mustang.

- cut costs and management layers, and
- enhance the firm's profitability

As it cannot be repeated too often, *costs matter.* Like quality, expenses should not be allowed to run away. Sharp businessmen have seen that and, day-in/day-out they have used a sharp knife to cut costs. Sam Walton, the founder of Wal-Mart, had a principle: "Never buy somebody else's inefficiency" (by buying his high-cost goodies).

Walton's qualitative benchmarks were that administrative costs, taken together, should never exceed two percent of business, and expenses associated with business trips should never exceed one percent of purchases done. These principles should be etched in a plate that finds itself at the desk of every chairman, president, or CEO, as well as of every prime minister and every minister of finance.

Closely associated and also requiring hard work to be sustained in the longer run is quality. Dr. W. Edwards Deming* was the first to publicly support the thesis that the system of mass-production, which propelled American firms to worldwide domination in the 1950s, was hugely wasteful of

- talent,
- materials, and
- labor

His solution was to personalize work by involving workers more closely in production and quality of produce. He divided labor into teams, granted each team responsibility for the quality of what it made, and instructed it to check continuously for flaws in the goods. The underlying principle has been that of focusing on a narrower but well-defined target in order to obtain commendable results at production level. The emphasis is on results.

## 2.6 THE RISK OF MISMANAGEMENT

"Many businessmen are always establishing new beachheads," said Peter Drucker, "but they never ask: Is there a beach to that beachhead?" Planning

* Deming was mostly ignored in America. But postwar Japan followed closely his ideas, which rapidly became central to lean production, just-in-time delivery, and team-based work. More on Deming's work in Chapter 11.

for something that makes no sense is one of the worst types of mismanagement—and of reasons why companies go against the wall, mismanagement is the most frequent.

Table 2.2 presents the findings of a research project by the Bank of England on why banks fail. Twenty-two different cases, some involving well-known institutions, were analyzed and of the six top reasons behind bankruptcies of credit institutions, mismanagement has been by far at No. 1 position followed by poor assets.

There are many reasons characterizing what lies behind mismanagement, and incompetence is only one of them. People rise to the level of their incompetence, suggests the Peter Principle. Infectious greed is another (more on this later). The following list shows the 10 more frequent reasons I have found in my professional work why companies miss their opportunity and end in the dust of history:

- Confused or nonexistent objectives
- No grand design of the company's operations—just "do something"
- Long delays in reaching urgent top management decisions
- Ignorance about the market drives and the competition
- Business plans detached from reality
- "Me-tooism" rather than a factual study of competitive advantages
- Throwing money at the problem, hoping that it is solved
- Inadequate preparation matched by failing skills
- Don't touch the fiefs of "colleagues"
- Lack of internal control and of corrective action

Each one and all of these reasons constitute awful examples on mismanagement of corporate expectations—typically ending in disaster. The signs of going in the wrong direction are relatively easy to perceive in the normal course of business, but rubber stamp board of directors (and quite often greedy investors) look the other way and the company continues accumulating wrong way risk.

A CEO who promises year-after-year double-digit increases in profits is, in all probability, one willing to cut corners. Several of them did so in the banking industry during the first decade of this century and brought the U.S. economy to a crunch (the European economy, too). When the market rises, investors are willing to believe anything; when it falls, they are

**TABLE 2.2**

Why Banks Fail: Findings of a Research Project by the Bank of England

| Identifier of Institution | Mismanagement | Poor Assets | Liquidity Problems | Secrecy and Fraud | Faulty Structure | Dealing with Losses |
|---|---|---|---|---|---|---|
| I | X | | | X | | X |
| II | X | | | | X | |
| III | X | X | | X | X | |
| IV | X | X | | X | | |
| V | X | | | X | | X |
| VI | X | X | | | | |
| VII | X | X | X | X | | |
| VIII | | | X | | X | |
| IX | X | X | X | | | |
| X | X | X | X | | | |
| XI | X | X | X | X | | |
| XII | | | X | | | |
| XIII | X | X | X | | | |
| XIV | X | X | X | | | |
| XV | X | X | | | | |
| XVI | X | X | | | | |
| XVII | X | X | | | | |
| XVIII | | X | X | | | |
| XIX | X | X | | | | |
| XX | X | X | | | | |
| XXI | | | | X | X | |
| XXII | X | X | | | | |

prepared to trust nothing, but it is too late. Mismanagement is a collective act. The wrongdoers include

- board members and executives characterized by dereliction of duties
- executives who put their short-term enrichment through king-size bonuses and stock options, ahead of their responsibilities to their company
- auditors who become the servants of managers rather than of shareholders
- bankers who cut deals with dubious corporate partners and help hide their losses, and
- brokers promoting their own interests rather than offering dispassionate investment advice to investors

"Creative accounting" is one of the means used to hide mismanagement's damages till the day of crisis. It fabricates a make-believe world of fake profits that are used to generate unjustified incentive payments. Another gimmick is custom-made derivative products that turn losses into huge profits in financial statements, and continue doing so till the scandal breaks out.

The 2007–2010 deep economic crises revealed that during the go-go years that preceded it overaggressive CEOs and CFOs booked sales before getting paid, manipulated the letter of contracts they entered into, and inflated revenues. Many companies used derivative financial instruments for gearing, and every sort of dubious business practice found a home in unreliable company books.

One of the important lessons that we have learned from the most recent economic, financial, and banking crisis is that there has been a common thread connecting the different events that turned sour. The worst problems were created by poor governance—made worse by lax regulation which

- failed to check high leverage
- paid no attention to illiquidity
- allowed capital adequacy to be manipulated
- failed to control risk management quality, and
- applied no penalties for executive misconduct

As this list documents, the risk of mismanagement is ever present. An additional reason why management skills have been wanting is that many of today's CEOs lack crisis experience. They are younger than they historically used to be, which is a positive development but which also has the downside that they lack the training necessary to hear the ringing alarm bells.

These CEOs have gained most of their experience during the 1990s, a period of record economic expansion, and therefore know precious little about when and how to appreciate that there are limits to growth or what it takes to handle a crisis. In other terms, battle-hardened champions who had steered their businesses through ups and downs had been succeeded by "greener" people who engaged in the economy's 2004–2005 superleverage and its treacherous waters of a deep recession.

- Experience on how to deal with adversity becomes so much more important when the economy is down and everybody seems to be waiting for a lift that does not come.
- Experience is best translated in hindsight and foresight, which is very useful in the prevailing uncertainties with daily surprises characterizing the reaction of markets under stress.

Moreover, surveys are demonstrating that today's CEOs are more status-conscious than their predecessors. Michael Milken said in the late 1980s, in response to a query regarding his title at Drexel Burnham Lambert: "I am some kind of a vice president. I am not sure exactly whether I am senior or executive or first. It keeps changing."* Milken was more interested in deal-making than in status.

"Greener" and status-conscious CEOs are *reactive* rather than *proactive*. One of the characteristics of "greener" people is that they cannot wait. Every day they search the Wall Street financials hoping that the economy will get better. But during 2008 and 2009 it refused to improve, sliding further downward. In the meantime highly indebted companies sank into more red ink. And so did highly indebted countries.

Denis Healey, the former British Chancellor of the Exchequer, is credited with what has become known as "The Law of Holes." It states that "When you are in one, stop digging." The CEOs of a number of companies, particularly of big banks, have probably not heard of it. They dragged down their organizations by huge debt load and a management that had lost its sense of direction. This was mismanagement at its best.

---

* Connie Bruck, *The Predators Ball*, New York, Penguin, 1988.

Dr. Vittorio Vaccari, the Italian economist and philosopher, foresaw this course of events when, in 1990, he described "The frenzied pace of contemporary living, the fleeting nature of intentions and commitments, the slavery to consumerism, the indulging tolerance of permissiveness, the capricious adoring of fashion, the idolization of the unnecessary, the celebration of vice...all these mark a wound of contemporary humanity."* All these, Vaccari suggested, "act like an anesthetic, producing a state of moral numbness. And how rare it is that those who have fallen under the spell...wake up." The awakening came with the 2007–2010 crisis, when it was finally discovered that somehow ethics had taken their leave, even if everyone seemed to know that *if* we built a moral society *then* most people would like to be honorable.

* Operare, July–September 1990, Rome.

# 3

# *Functions of Management*

## 3.1 SIX BASIC FUNCTIONS

Management exercises its art through specific well-defined and disciplined functions. Although no two management practices are the same and every company tends to have its own characteristics on how it runs its business, management functions can be grouped into six classes as shown in Figure 3.1. These range

- from the unstructured environment of forecasting, where exist many unknowns and degrees of freedom
- to the structured environment of internal control, where the rules should be firm and clear while the system is working like a clock

Theoretically, *forecasting* (Chapter 4) is concerned with the prognostication of future events and happenings. Practically, as it has been already brought to the reader's attention, the most important contribution of forecasting is that of evaluating the future impact of current decisions—which is a demanding task requiring

- foresight
- experience, and
- a great deal of objectivity

*Planning* is another basic management function. It is an intellect-intensive process aimed at the determination of a course of action down to the detail of the resources being required. Corporate planners base their proposals to top management on objectives, forecasts,

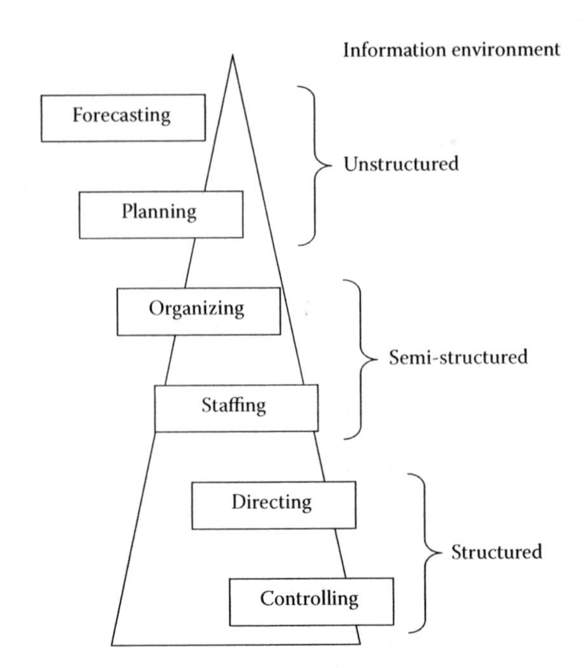


**FIGURE 3.1**
The functions of management can be grouped into six classes with distinct characteristics.

available assets, expected returns, possible implications, and projected requirements for corrective action.

Such proposals must be written in a clear and comprehensive form to be confirmed or changed by top management. They should also encourage the greatest standard of effectiveness in execution. Plans must account for people who are able to carry them out because, as it cannot be repeated too often, the first syllable of management is *man*. The best plans

- target attainable, but not trivial goals
- take account of human resources and other assets, and
- prescribe a disciplined movement toward reaching sought-out objectives

In this, corporate plans resemble military plans, prepared by the general staff. The function of disciplined movement in battle is to produce in the mind of friends the conviction that they cannot lose, and in the mind of foes that they cannot win.

Plenty of examples help in documenting that effective planning premises will heed the advice given by H. Ross Perot* in his principles of

* Of EDS and General Motors fame—in the latter case as prognosticator that poor management will bring GM to bankruptcy. It did.

management: Every good thing stands moment-by-moment on the razor's edge of change. It must be fought for, to be retained. Contrary to what long-haired theoreticians suggest, Perot believes that the entrepreneur is a proponent of

- Push, push, push
- Harder, harder, harder*

That's the *doer's* nature. Perot also brings to the attention of those who listen to his advice that business is not a social service—and he advises people who ask for "security": "You are your own job security."

"A company must always have a single vision and a common business plan, for without this guiding focus, it will be directionless," say C. Gordon Bell and John E. McNamara. "Although such a plan can be changed, one and only one plan must exist at any given time, and everyone in the organization must be trying as hard as possible to carry out that plan."†

To meet its plans in an able manner, the entity needs *organization and structure* (Chapter 5). The human resources of a company are not being put together like a little hill made of grains of sand. A firm's know-how as well as its financial and physical assets must be managed through well-defined organizational units with authority to do so—and responsibility for deliverables.

Organization and structure change over time to adapt to the redefinition of business and to match changing market conditions, requirements posed by innovative products, and moves by competitors. But at any given time there must be a line of command that permits to move the entity forward, and makes feasible feedbacks, which serve in adjusting the company's course of action and in cases its structure.

The plan impacts upon the organization and, as we have just seen, the structure serves many purposes, one of the most important being that of indicating where human capital should be raised to meet the challenges ahead. The company's ability to be up to its vision, its market image, and trajectory will be conditioned by the quality of its human

---

* And he is an irritant to the bureaucracy, because he stirs things up, says Perot.

† C. Gordon Bell and John E. McNamara, *High-Tech Ventures*, Addisson-Wesley, Reading, MA, 1991.

assets—all the way from selecting to steady training. That's the remit of *staffing* (Chapter 7).

The functions of staffing range well beyond the classical hiring, providing offices, and budgeting personnel expenditures. Human resources must be animated and led (Chapter 2). They must also be steadily upkeeping their know-how, because skills decay. In the late 1980s a study by the Bank of Wachovia documented that a banker who does not get training for five years loses 50 percent of his skill. Today, that pace has quickened.

Staffs become depressed when confronted by the waste, incompetence, and downright obstructionism of some of the people working for the firm—particularly so if these happen to be managers. The mission of *directing* (Chapter 7) is to see that things are happening according to plan through able management of human resources.

Within the function of directing fall as well many *tactical* decisions that have been taken in a timely manner to honor the mission assuring that things go according to plan. George Anders provides an example from KKR, the investment firm. The theme is KKR's exploitation of tax breaks associated to debt and leverage.

> Big tax breaks came from loading up a company with debt and boosting its depreciation deductions.* Tax considerations play an important role in increasing companies' cash flow after buyout; KKR confided to potential investors in a 1982 fund-raising memo. At RJR Nabisco, for example, the company's tax bill shrunk to just $60 million in the second year of KKR's buyout, down from $893 million the year before KKR acquired the company.†

Yes, but after the LBO Nabisco has never again been able to stand on its feet.

Control is the sixth basic function of management. An organization will run wild without a strong function of *internal control* (Chapter 7). A sound system of internal control is synonymous with taking the proverbial long, hard look at how the business should be planned, organized,

* See Chapter 16 on the ills of leverage. Some textbooks suggest that "modern people" leverage themselves. I would rather think that superleverage is done by speculators and by those other people who are stupid.

† George Anders, *Merchants of Debt*, Basic Books, New York, 1992.

staffed, directed, and controlled so that the right things happen at the right time—and deviations are put straight before it is too late.*

Because organizations are made of people, the effectiveness of management control depends a great deal on understanding the business and its people. "Problems arise when people at the top don't understand the professionals working for them, and therefore they can neither guide them nor control them," said Brandon Davies in a personal meeting.†

The six basic functions of management briefly described in this section are analyzed and detailed in Chapters 4, 5, 6, and 7. Prior to this, however, it is necessary to review other functions like policy-making and the management of change. In the longer run the way these are being performed distinguishes a well-governed organization from one going down the road to oblivion.

## 3.2 CORPORATE POLICY

*Policy* is a guide for thinking and for avoiding having to make time and again the same decisions. It is a mental framework on which will be built the process of longer-term, medium-term, and day-to-day management. Clearly established company policies help in eliminating day-to-day decisions on predicted subjects—but to be effective

- they have to be kept dynamic, and
- they must be constantly reevaluated.

Like strategies, policies are nothing to set once and then forget about them. The same is true of plans. "One does not plan and then try to make the circumstances fit those plans," said General George Patton. "One tries to make plans fit the circumstances. I think *the difference between success and failure* in high command depends on the ability, or lack of it, to do just that." Patton's statement is equally valid with policy-setting.

---

* D. N. Chorafas, *Implementing and Auditing the Internal Control System*, Macmillan, London, 2001.

† Davies was at the time the Treasurer of Barclays Bank.

The art of policy-making is to elaborate them in such a way that they are both firm and adaptable, and also open to critical review and change when they become obsolete or have proven wrong. Sam Walton, one of the most successful businessmen of the post–World War II years, described in the following manner the way his mind worked: "When I decide that I am wrong, I am ready to move to something else."* A policy may prove unable to deliver, and as Walton aptly remarked, it has to be changed.

Examples of policy decisions are the firm's product line, product mix, customer handling, innovation, cost control, leverage, make-or-buy, inventories-to-sales ratio, hiring and firing practices, discounts, marketing thrust, the use of high technology in management information, and more. The active search for partnerships as well as mergers and acquisitions are other examples of important policies.

When, in the 1960s, Tex Thornton, the CEO and promoter of Litton Industries, was asked by journalists why his firm was bent on acquisitions he answered "We don't buy companies, we buy time." When today pharmaceutical or IT firms buy start-ups to capitalize on their products, they too buy time.

This being said, companies in the same industry may adopt diametrically opposed policies because of difference in business philosophy. A key policy question, for instance, is "Shall we 'make' or 'buy?'" Another, "Shall we be 'first' or 'second' in our industry?" Sometimes the answer is situational, and in other cases deliberate.

Back in the early 1950s, Remington Rand's purchase of Univac, after IBM's Thomas Watson Sr. refused to buy it (Chapter 13), led the former company to stellar performance in computers. But by the early 1960s IBM caught up with and left Univac behind. The adopted policy made the difference.

IBM's policy was based on two pillars. The one was intensive management attention on R&D for fairly rapid internal hardware and software developments. The other was extrovert, calling for an intensive strategy of client education on computer usage—and it was a "first" in data processing. The two together permitted IBM to gain time in order to overtake Remington Rand Univac.

Policies are typically established by the board, CEO, and his immediate assistants. But they are not necessarily communicated to a wider audience. Most often, senior managers keep them close to their chests. At the same time, however, as formal statements giving the relationship between plans,

* Sam Walton, *Made in America: My Story*, Bantam, New York, 1993.

decisions, and action policies may become obsolete without steady and timely reevaluations.

Critical reevaluations of existing policies should be made in full appreciation of the fact that to be vital to the enterprise policies must be current—even if by being general statements they are usually not too specific. In addition, to channel their decisions and actions in accordance with corporate policies subordinates must *interpret policy.*

The notions described by the previous paragraph constitute at the same time a policy's strength and weakness. While a policy's strict implementation permits the coordination of all elements into the general framework, there exist prerequisites at operative level such as

- guidelines for interpretation of policies
- criteria for tactical moves, and
- the exercise of a rather uniform ground judgment within a given pattern of activity

In addition, a well-thought-out longer-term management policy may change the company's pattern of activity. Siemens, the German electronics and electromechanical manufacturer presents an example. As Figure 3.2 shows, in 1975 Siemens derived three quarters of its income from mechanical and electrical engineering, and the balance from electronics. A focused management policy saw to it that

- in 10 years time, the company's income came 50–50 from M&E electronics, and
- ten years later, electronics represented three quarters of the business (This share further increased during the ensuing years.)

Risk management provides another example on policy impact. During the go-go years 2005 to mid-2007, Goldman Sachs policy was much more open than that of rivals to take risks, but it also was quicker to hedge them. For instance, in late 2006 it spent up to $150 million, corresponding to one-eighth of that quarter's operating profit, hedging exposure to AIG.

Goldman had the policy of promoting senior traders to risk positions, letting it be known that such moves are a potential stepping stone to the top. These traders turned risk managers were encouraged at hedging tail risks. (Tail risks are those at the long leg of a distribution that are typically low to very low frequency but high to very high impact.)

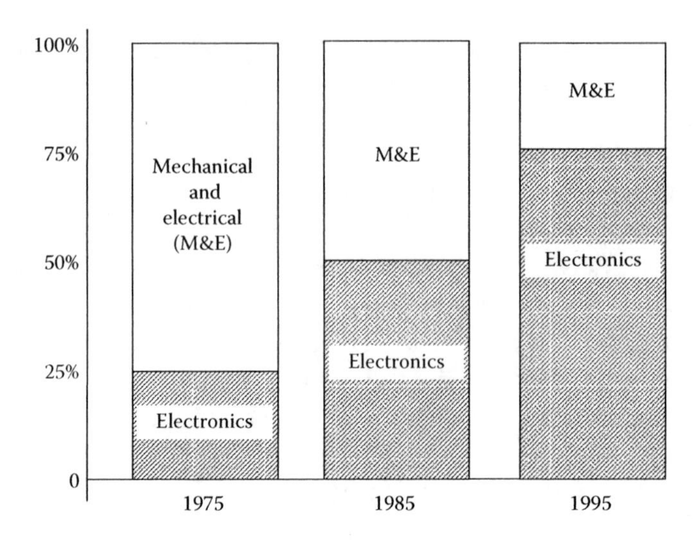


**FIGURE 3.2**
Evolution in yearly business by Siemens over a 20-year timeframe.

By contrast, in other big banks that went to the wall during the severe 2007–2010 crisis had a high-risk appetite but the policy governing their risk culture was substandard. Their executives were free to bet with subsidized funds, while transfers from the bank's treasury were deeply underpriced and assumed exposure was presented in a "net and forget" format. For instance:

- trading desks estimated the maximum possible loss on risky assets, and
- they hedged it but they also recorded the net risk as minimal, concealing huge trail risks in the gross exposure

To make matters worse, in spite of substandard risk control, management policy encouraged risk-taking, rewarding it with hefty bonuses. Failure to promote risk control, became a corporate culture and banks ran downhill, wounding themselves badly and coming to the edge of bankruptcy—to be saved by a lavish amount of taxpayers' money. Examples are Citigroup, Bank of America, and UBS. Others fell in the abyss like Lehman Brothers.

- Excessive risk-taking and absence of rigorous risk management have been policy faults.
- Substandard risk controls and clogged internal control communications lines, were errors associated to procedures.

*Procedures* must not be confused either with policies or with plans. They are means of action; therefore more detailed than policies and better controllable. Within, say, a profit-planning system, procedures should bring forward the profit and cost elements, the latter in connection to budgeted expenses. (Cost of sales, manufacturing, and R&D costs; profit appraisals; inventories; turnover, available liquidity, and more.)

Procedural solutions may prescribe tools like opportunity analysis; for instance, in connection to inventory problems. They may focus on assignment of resources, transportation, scheduling, and other themes in which management has alternative possibilities for decisions. Procedures, for instance, may specify the use of models with optimizing/allocating algorithms or require dissenting opinions in matters concerning money, time, goods, machines, people, facilities, raw materials, and the like.

Procedures established by sophisticated organizations press the point that experimentation is essential to management planning. Through opportunity simulation we can express the relationship prevailing within a system in quantitative terms, but procedures may also specify that the model should present as nearly as possible the real life environment being emulated—and it should be tested in depth before being implemented.

## 3.3 THE MANAGEMENT OF CHANGE

One of the most challenging issues confronting senior management and the functions it is expected to perform is the management of change. Typically organizations are suspicious of change. The risk being perceived is not so much financial but one of status quo and acquired benefits—because power and prestige can be lost. Therefore, the management of change

- starts with the marketing of change
- requires a well-conceived plan and timetable
- leads to a new structure, and
- aims at a quantum leap in efficiency

The marketing of change requires both leadership and profound knowledge of the people in the organization, as well as of the tasks they perform. This involves not only the understanding of corporate problems, but also the appreciation of practical ways to address them. It also raises complex issues such as changes in managerial thoughts and practices. It also requires problem-solving skills, since change unavoidably involves several internal and external problems.

At the time of Thomas Watson Sr., structural change at IBM was common currency. The way a saying at the time went, the company's hierarchy was like a hill of sand. Every three to five years God was pressing his finger in that hill and moving it. So

- the grains at the top found themselves at the bottom, and
- then, as God used his palms to reconstruct the hill, grains that were at the bottom rose to the top

Structural change obliges to reconceptualize management's functions; critically analyze how the rising curve of innovation and globalization affects the firm's clients, its people, its product(s), and its market(s); and pay much greater attention to planning and training—which is part of rethinking one's mission. The company's people who intend to survive, must be prepared to reinvent themselves by

- thinking the unthinkable
- challenging the obvious
- adjusting swiftly to the switches, and
- de-constructing increased maneuverability

Thinking the unthinkable requires testing new ideas about products and processes; developing new market opportunities in domains that in the past were off-grounds; involving the best people at all levels of the organization, preferably starting at the top. It also makes mandatory to use high technology (to leave one's competitors behind), and look at all cost items like a hawk.

Moving fast in managing change is a major asset. There have always been exceptional individuals, and companies, capable of moving fast seeing through their policies without allowing slacks to develop. After salvaging Turkey from disintegration, Mustafa Kemal Atatürk favored replacing Arabic with Latin script. As it can be expected, this raised fierce resistance.

Mustafa Kemal confronted it head-on. Once he made up his mind there was no letting go. Steady pressure was exercised not only at the top but also at the bottom. He went to towns and villages and talked to the common man. Then, once he got the people engaged he saw to it that reform was carried through within six months.*

This drive to get results in spite of mounting resistance, or more precisely because of it, is not the way the average executive operates—or the typical organization reacts. Change has its detractors and its enemies. The forces against change feed on

- people with decision power who are afraid to lose their authority, their might, and may be their job

This leads to internal politics and "resistance fronts," which sometimes succeed in freezing change in the organization.

- The concepts that prevailed 20 and 30 years earlier, which still dominate thinking and product design.

For example, IBM's instance in the mid- to late-1980s to base its fortunes on mainframes has been an unmitigated disaster. In those years, PCs were sweeping the market, client/servers were much more cost-effective, and database computers were gaining market share.

- The nature of key criteria for an epoch full of changes in technology had not been elaborated.

Part of the reason was a massive resistance to change among the corporation's top management. The status quo was king and bureaucracy has a field day. Resistance to change is easier to overcome when the criteria for performance, and reasons underpinning them, are not only properly established but also explained to all stakeholders. The best approach is to convince that both they and the firm have much to lose by opting for the status quo. The rationale for change is best explained by identifying

- which has been the point of departure,
- how did management arrive to the conclusion that living in the past is harmful, and
- which are the rigorous links between cause and effect leading to the conclusion that change is the best possible course

* Andrew Mango, *Atatürk*, The Overlook Press, New York, 2000.

Failure to fulfill what is asked by the third bullet is largely a failure in forecasting and planning (Chapter 4), having its roots in "thinking as usual." Also, in believing that the mountain will go to the prophet, rather than vice versa. IBM nearly fell to the abyss because of that hard-headed policy.

Eventually, resistance to change ebbs and there are plenty of examples to document that in the end it has been futile. *Luddites* provide an example. The original antichange, antimachine revolutionaries known as Luddites, were active in the industrial countries of England for 15 months in 1811–1812. They took their name from Ned Lüdd, who allegedly smashed a textile frame in Leicestershire.

Working masked and mainly at night, the Luddites are remembered for having destroyed the mechanical looms that were taking away their weavers' jobs. Eventually the British government dispatched 14,000 troops to quell the industrial sabotage and the Luddites were crushed. About 15 of them were killed during the protests, 24 were hanged after a trial at York, 37 were shipped to Australia, and 24 others were imprisoned.

But not everything was in vain. Because of this shocking upheaval, in the following years, improvements were made in wages and conditions for textile workers. With labor peace reestablished, the British textile industry saw more machines than ever before. In memory of that armed resistance to change, Luddism now means any futile negation of progress. The term is also linked to the fallacy in economics that the demand for labor is fixed in the short run and labor saving machinery kills jobs.

## 3.4 RESPONSIBILITIES COMMENSURATE WITH AUTHORITY

It is an old management maxim that only authority can be delegated. Responsibility is never delegated. It is accepted when one assumes the authority to do something, but it also stays with the person higher up in the organization chart who did the delegating.

In this sense the CEO is responsible for the accuracy of financial reporting to shareholders and supervisory authorities, even if this function depends on the CFO. But the CFO, too, holds integral responsibility of what is written in the books and in financial reports though the person directly responsible for the company's books is the chief accountant—who is, evidently, coresponsible.

There is no better example on this than the Sarbanes–Oxley Act (SOX). In 2002, with the uproar of the Enron scandal, the U.S. Congress voted a law aimed at redressing the financial reporting practices of American corporations. Its further out objective has been to improve corporate governance by providing more accurate financial information to investors. The SOX Act

- held CEOs and CFOs directly responsible for their financial reporting, and
- reinforced the independence of the audit system, but strengthened the importance of reports by independent public accountants

Resistance to change, however, showed up not only among the companies but also at top government level. To be properly implemented, SOX required to reinforce the powers of the Securities and Exchange Commission (SEC), barring executives who abuse their power from serving in other corporate jobs. But George W. Bush decapitated the SEC by firing William Donaldson,* and the implementation of SOX became spotty at best.

In spite of the letter of the law, investors did not get access to more dependable information necessary to judge a company's financial condition and risks it had been assuming. Neither did they benefit from *prompt* access to critical information. The 2007–2010 debacle has documented how wrong has been this blind-eye policy by the government and by regulators.

SOX stated that chief executives are personally responsible for the accuracy of their companies' public disclosures and financial statements. In the ensuing years, since 2002, plenty of false financial statements have been filed. The latest to hit the news (in March 2010) is that prior to the investment bank's bankruptcy Lehman Brothers' top brass hid some $50 billion in losses. But nobody, just nobody, has been brought to justice.

SOX also stipulated, between the lines, that company executives will be banned from profiting from false financial statements. Nothing has, as yet, happened to really apply the letter of the law. Instead, huge fortunes were made out of public misfortunes.

Other rules that have become laws stipulated that independent auditors (certified public accountants [CPAs]) should not carry out other services

* Albeit on the stated reason of registration of hedge funds, to which the Bush administration was *irrationally* opposed.

for the entity they audit, if these services compromise the independence of the audit. Up to a point, this has been done, as the big CPA partnership sold or spun off their data processing consultancies. But at the same time they developed other paid-for lucrative services marketed to the companies being audited.

The permeability of top positions between CPAs and their client firms, is also a harbinger of ethical problems. People who want the capitalist system to survive by reinventing and restructuring itself, suggest as one of the measures a 5-year waiting period during which auditors may not accept senior positions with audit clients. Other suggestions center on forbidding creative accounting practices, doing away with top management's laxity, as well as

- increasing disclosure of assumed risks and assumptions built into financial statements
- establishing harsh penalties for executives, directors, and auditors who fail to properly discharge their responsibilities, and
- barring corporations from hiring their audit firms to do nonaudit and nontax-related consulting

SOX moreover stipulated that a regulatory board will assure the accountancy profession is held to highest possible standards. This board, known as AICPA, has been put in place. The problem is that nothing came out of it with positive impact on

- greater accuracy of financial reporting, and
- certified public accountant ethics, standards, and practices

Instead, vague statements have been made, such as "company accounting systems must be characterized by the best possible practice"—which mean zero-point-zero in practical terms. Critics suggest, not without reason, that a major weakness in applying the letter of the law is that politicians play havoc with it.

Critics of what is wrong with the fact that nowadays responsibilities are not commensurate with authority, say that when one is candidate for an election he or she expresses different (often contradictory) opinions because voters are not characterized by a unique state of mind. In a way, the more contradictory opinions the politician has, the greater the chances he gets elected.

Even after one is elected, contradictory opinions can be of help to one's political career, particularly when expressed as deep rooted. When in 1925 the government of Edward Herriot fell in France, an article made this commentary: "Whichever may be one's opinion about Herriot's government, it is proper to recognize that he respected the essential issues of his programs. He respected them so much that he did not even touch them."*

That's precisely what has happened to SOX in two consecutive administrations. Implicit to the SOX Act was that company executives should go to jail for cheating shareholders. The 2007–2010 deep economic crisis unveiled plenty of cases for bringing to justice wrongdoers, but this has not happened.†

Yet, by asking CEOs and CFOs to personally certify the accuracy of their companies' financial statements, the law of the land made it easier for the government to punish wrongdoing executives. It also aimed to reduce the ability of officers and directors to insure themselves against fraud claims. This, too, however, was put to the backburner as companies continued being able to indemnify directors and officers against fraud lawsuits.

Critics of the current very light regulation and supervision want to see tougher discipline for officers, directors, and auditors who seriously fall down on their responsibilities. When executives willfully mislead the public, there should be penalties stiff enough to instill a deep sense of accountability. At the time of Enron's dishonest practices, *Business Week* had suggested that there should be legislative measures to protect whistleblowers like Enron's Sherron S. Watkins.‡ Nothing like that has happened and this led to the Madoff Ponzi game.

It should be recalled for the sake of good order that at the time of Enron's collapse its employees lost both their jobs and their pensions because the company's pension plans held high levels of the entity's stock which, under the plan's rules, workers were not permitted to sell when the price fell. By contrast, senior Enron executives sold millions of dollars of the company's shares and options before the price collapsed.

In 2006 Jeffrey Skilling, Enron's former chief executive, was sentenced to more than 24 years in prison and ordered to pay $45 million in restitution for his part in the fraud that bankrupted the energy trader in 2001.

---

* Jean Egen, *Messieurs du Canard*, Stock, Paris, 1973.

† Ironically, in early 2010 Dennis Kozlowski, Tyco's CEO condemned and imprisoned for fraud (albeit not because of SOX) even asked for compensation for lost executive privileges.

‡ *Business Week*, March 11, 2002.

But the convictions of Kenneth Lay, who was the chairman and promoter, were vacated following his death in July of that year.

As far as executive responsibilities are concerned, also a "must" is legislation, which assures that corporate leaders are required to disclose trading in their companies' shares within two days and that directors who abuse their positions of trust should lose their right to serve on boards. Short of reinforcing the ethical side of senior executive's responsibilities, hopes of playing by the rules will be half-baked.

## 3.5 MANAGEMENT BY OBJECTIVES

The term "management by objectives" originally came up as a label to sell consulting services. But there is sense to it. In fact, the practice preceded the label. In the 1960s, when I was consultant to Brainard Fancher, the CEO of General Electric-Bull, every year in early December he was asking the vice presidents reporting to him to

- present in writing their 10 top objectives for the following year, and
- a year later, just prior to Christmas, to report also in writing on how the, by then 12-month old, objectives were met

Behind this approach lies the fact that a set of properly stated 8 to 12 annual objectives help in developing a dynamic *strategic framework*. The prerequisite is that such annual objectives are realistic and written in the light of identified opportunities. Technological, financial, and marketing goals can be examined under the dual aspect of market forces and vital company operations.

Strategic reviews, top management choices, and day-to-day tactical decisions can be helped by the *entification* of crucial factors, made feasible through well-established objectives. However, in order to serve in this way, objectives must

- be meaningful, precise and tangible
- be set in a workable time span with realistic target dates, and
- offer benchmarks for measuring progress as well as for exercising corrective action

As the foregoing references suggest, management by objectives can be developed into a mechanism providing the spirit and means for effective planning and control of the organization. This requires a methodology as well as plenty of practice in applying, well beyond the aforementioned task of setting objectives. The manager must

- identify the activities needed to complete each project under his control, and isolate those that are critical
- project resource requirements, by type of resource and by time period
- establish schedules, summarize status, and highlight those issues that require attention, and
- elaborate a control mechanism with walkthroughs, allowing to analyze failures and discern reason(s) as well as responsibilities

Not everybody appreciates that failures and errors can be turned into golden opportunity for learning. To learn from failures, we should develop and steadily update a list of subjects that can go wrong, review it regularly, identify the reasons—and, when something does go wrong, provide corrective actions. Only then can we have a practical effect on the origin(s) of failures, which stop us from meeting objectives.

It is no less true that this procedure can have a much broader applicability, because it starts with a plan to identify all of the activities required to complete our *mission*. It also includes the manner in which these activities are related, the estimated duration for each, and the responsible individuals who should give their undivided attention to the mission's accomplishment.

The next step in management by objectives is *schedule preparation* aimed to determine the estimated duration, earliest possible starting and completion dates, and latest starting and completion dates that will not delay or interfere with the mission's accomplishment. This should be followed, as a third step, by a *hot list* that brings to attention the most important items for each manager. "Hot lists" identify the activities

- critical to completion of the mission, and
- scheduled during the next several weeks, thereby providing the ability to look forward

The fourth step in management by objectives involves updating. Information must be collected regularly on activities completed since the previous scheduled check, those started but still not completed, new activities to be added to the schedule, and changes in activity duration, as well as in responsibility or description.

*If* there is full agreement on objectives, *then* coordination is easier. If there is agreement on the results of organized activity, there is little necessity for prior approval of every minor operational decision. The scene is then set for delegation of operational authority, with senior management reserving to itself

- the final approval of objectives, and
- a review of performance through accountability reviews

Objectives can moreover be used for training personnel to think in terms of the organization as a whole and to provide motivation for individual initiative, including the resolution of problems raised by innovation. For example, borderline issues between research and marketing, bringing together

- R&D plans
- marketing investigations
- financial performance estimates, and
- strategic evaluation for each product or service

From statements made in the preceding paragraphs, the reader will appreciate that management by objectives has prerequisites. It is not just a "handy term." Truman's and Napoleon's concept on the need for follow-up on the execution of an order (Chapter 1) is one of them. An interactive *key activities* report should be steadily updated to identify—milestone by milestone—schedule completion dates. The manager needs this report in his review of progress to

- apply a coherent project control authority
- evaluate not only overall schedules but also details in execution
- translate developing "needs" into specific plan revisions, if necessary
- request specific support for activities under his command, and
- generally enhance project planning and control capabilities

Changes may be necessary to correct organizational inefficiencies, improve morale amongst members of management, upgrade specialist groups, or revamp career planning concepts and procedures. In the process of correcting organizational inefficiencies individual executives should be made responsible for the accomplishment of each task within explicitly stated deadlines. And as already stated, steps should be taken to provide for regular reviews of progress toward objectives.

Management by objectives as well provides a testbed for policy evaluations, and for countermeasures to meet business maneuvers by corporate opponents. Prerequisite to this is the definition of objectives both qualitatively and quantitatively—in relation to markets, market share, return-on-equity, return-on-investment, and other road maps or milestones to be reached. For instance

- manpower evaluation, including productivity, quality of deliverables, hiring, and training
- direction of research to meet product strategies
- manufacturing potential, including capital investments (capex)
- marketing and sales capabilities for market leadership, and
- financial evaluations, including budgets, liquidity and cash flow (Chapter 14; also standard costs [Chapter 15])

Depending on the accuracy and timeliness with which objectives are met, corrective action may be necessary to ensure that effectiveness in research, manufacturing, marketing, and sales is improved. The emphasis must be both operational and economic with profitability a pivot point. Sales targets must be set, cost and expense budgets fixed, and policies reviewed and, if necessary, rethought and reestablished.

## 3.6 MANAGEMENT BY RESULTS

A manager is judged by the results he delivers, and not by the clapping of his supporters or the whistling of his opponents. The results are shown through deliverables that are measurable and (save exceptional cases of an unanticipated opportunity) conforming to plans.

A good manager would allow himself to be exhausted but never to be defeated. In counterpart, he will be often tired but never bored. That's the advice I got from Jim Allen, the man who built the Booz, Allen & Hamilton consultancy, when in 1960 I joined his company. Boredom, not fatigue, is the enemy of life.

Five long decades of experience have taught me that Allen's words were wisdom. He also had said that *management by results* is the best and fairest way to evaluate a consultant or any other professional at any level in the organization. At GE, Jack Welch, the company's long-standing CEO, held annual "Session C" meetings during which he

- personally evaluated the performance of GE's top several hundred managers, and
- funded "imagination breakthrough" projects that permitted his managers and professionals to think out of the box

These meetings included general managers of GE's major business units, as well as senior engineering, manufacturing, sales, marketing, and financial executives. The goal was to better appreciate these peoples' deliverables and to extend GE's boundaries by

- evaluating careers
- priming cost-cutting and efficiency, and
- linking bonuses to new ideas, customer satisfaction, and sales growth

Every company will be well advised to follow this management by results policy because it is a solid way for assuring that *our* company's leadership is effective, and the firm itself is a high-quality, low-cost producer of innovative products with market appeal. This is true of every functional area and of every project the organization undertakes.

Take IT as an example. It may sound ironic because computers and communications is by now a bread and butter business, but one of the weaknesses I repeatedly found in top management is its inability to guarantee that *technology* is used in an effective manner to give profits and keep the company at *cutting edge.*

- *If* successful CEOs are characterized by keeping their IT subordinates on the run,
- *Then*, based on long decades of experience, I must say there are not so many successful executives around.

When it comes to IT deliverables, a good deal of presidents, executive vice presidents, and their immediate assistants (including CIOs) will not pass the test of management by results. Some of the worse examples come from financial companies.

This may come as a surprise, because banks of all sorts, from credit institutions to brokers and investment banks, have spent more than any other industry on IT. The money they threw at IT during 2009 amounted to a wholesome $500 billion according to Gartner, a consultancy. In spite of that,

- the quality of information filtering through to senior managers is inadequate and often substandard, and
- while the hardware may be new and very expensive, the software is prehistoric, its sophistication limited, its quality questionable, and programmer productivity is abysmal

In October 2009, amidst the deep economic crisis, a report by bank supervisors pointed to poor risk "aggregation." Many large banks simply did not have the system solutions able to present an up-to-date picture of firm-wide links to borrowers and trading partners.*

The same reference pointed out that two-thirds of the banks surveyed said they were only "partially" able to aggregate their credit risks. This and other findings led to "discoveries" that shocked the examiners. Some of the big banks needed days to calculate their exposure to

- derivative instruments, and
- trading counterparties

Had the CEO, CFO, risk managers, and CIOs of these banks undergone a *management by results* test, they would have flunked it. It looked like the unable was being asked by the unwilling to do the unnecessary—while the necessary was not done.

True enough the calculation of derivatives exposure, including market risk and counterparty risk, is not easy. The instruments themselves are made unnecessarily complex and bilateral deals typically involve esoteric clauses. There are as well complications associated to legal matters in different jurisdictions. But

- when one does not know how to calculate the risk he or she is taking,
- then the most rational thing to do is not to assume it

The answer "I did not know" is not receivable. The people assuming such God-size exposures—with shareholder and taxpayer money, not with their

---

* *The Economist*, February 13, 2010.

own—are very highly paid executives and traders who should know what they do. They are not babies from the kindergarten.

- *If* they don't know how to compute and update the exposure they assume,
- *then* management by results is at its lowest; they don't qualify for the job they do.

Not unexpectedly, the banks with the most dysfunctional information systems like Citigroup, were the most hurt from the debacle. They were through multiple mergers and acquisitions subsequently failing to integrate the technology of the various firms under their wing. This, too, is a documentation that their CEO, CFOs, and CIOs have failed the test of management by results.

Voices at Wall Street have been saying that in the depth of the crisis some banks were unaware that different business units were marking the same assets at different prices, and that among themselves they had largely overrun credit limits. The way forward is not to reject high-tech finance but to be honest about its limitations, says Emmanuel Derman, professor at Columbia University.* In other terms,

- *If* CEOs and their pals don't know that there are limits,
- *then* they should be forbidden to practice to trade, lend, invest, and assume risk.

Shareholders and taxpayers, who among themselves food the bill, are not that stupid; more or less they know what is going on. Sometimes the reaction is full of folklore as it were in the week of March 16, 2009, when a crowd of shoe-throwing demonstrators gathered outside a conference center in Canada, as George W. Bush spoke to businessmen in his first speech since stepping down as U.S. president.†

Not only the executives and managers' deliverables should be subject to *results-testing*, but also the policies they establish and follow, because they are at the origin of inordinate and unnecessary costs. Some of these policies are a total aberration, adopted mainly to please the gallery, but ending by

- satisfying nobody, and
- costing a fortune in perpetuity

---

* Idem.

† *The Economist*, March 21, 2009.

An example is the European Commission's policy that its documents are translated into more than 20 languages. These translations are done by experts—and they cost a fortune. Nobody has bothered to establish a one-language rule whereby all people and entities having to do with the European Union (EU) will communicate in that language.

What is now happening in the EU is fools' paradise. Thales, one of the eight sages of Antiquity, used to say that what one fool can do another can do, too. Dr. Carlo Pesenti used to answer the complaints of CEOs of his banks that I was taking them for fools with a short reply: "If you don't want to be taken for a fool, then don't do foolish things."

Management by results is to a substantial extent a test of *foolishness*. Leniency toward a bad CEO, CFO, or other senior executive is not a policy that bears fruit. Whether the reason for getting the incumbent CEO out of the system is underperformance or malfeasance he has to be replaced—and the same is true of all executives. The board and shareholders should heed the advice of an Athenian senator in Shakespeare's *Timon of Athens*: "Nothing emboldens sin so much as mercy."

# Part Two

# Management Principles

# 4

# *Forecasting and Planning*

## 4.1 FORECASTING

*Forecasting* aims to project into the future; it is doing so either by foretelling coming events or by pre-analyzing the future impact of current decisions and acts. As the careful reader will recall, the latter is the main objective of business forecasting. It is, as well, the more successful of the two missions.

In a significant number of cases, this projection into the future is thought to be effective, but it is practically unavoidable that it is influenced by subjective judgment; and that it involves speculation. *The perfect forecast has not yet been invented.* But even though forecasts can never be as precise as a watch they can be valuable, particularly so *if*

- they reflect learned assumptions about specific cases, and
- they are tailored to fit the needs of the organization on whose behalf they are being made

General type forecasts often come out of a cuckoo cloud. By contrast, specific forecasts are valuable because they represent the best estimate that can be made about a future situation at the time of their preparation.

History books say that the wish to learn about what the future has in store finds its origin in antiquity. It was in the mind of dignitaries who flocked to the Temple of Delphi in ancient Greece bearing wholesome gifts in the hope to learn about future events.

More recently, in the post–World War II years, the "future" has become a topic of increasingly widespread concern and interest, but its study is not always done in a systematic and structured way. To a substantial extent it is based on guesstimates. Assumptions made in forecasts (Section 4.2) must be pragmatic, because no matter how well a company schedules its operations, its plans collapse if forecasts are seriously in error. Management must know in advance the answer to issues such as

- the *most likely* way in which the economy will develop
- the kind of market opportunities that will exist
- how competitors will react to *our* new products
- the quantity of sales that can be expected
- how fast costs will rise, and in which domain(s)
- whether or not new plants and sales offices will be required to serve customer needs

Forecasting does not aim to foretell such future events in high degree of accuracy. Its objective is to provide a basis for judgment, which serves to choose among alternative courses of action, and establish a roadmap that can provide guidance. This offers a background for corporate planning, as well as reevaluation of current policies.

Forecasting is a prerequisite for planning just like the setting of objectives is a prerequisite to the formulation of strategy. The benefit obtained from forecasting can be measured by the success managers and planners have in avoiding crises and gaps.

The research into the future has grown in importance, because technological superiority is essential to industrial leadership. The timeframe for projections and prognostications varies with their objectives. We broadly distinguish

- *long-range* forecasts, most often limited to between five and seven years, but can also go further out
- *medium-range forecasts*, typically at the two to three years level; they help in connecting long- and short-range plans, and
- *short-range forecasts* serving for budgeting, manufacturing, and sales projections, the latter typically made every six months for the next 12-month period

Within these classes exist further breakdowns. For instance, sales forecasts can be made by aggregates of products, known as "commodity groups." The forecasting of stock requirements for individual products per stocking period, is another short-range example—and a very important planning tool indeed.

Marketing forecasts for new and current products are typically made in cooperation with other division. New product sales forecasts coinvolve the product planning department. Design engineers work together with manufacturing experts in their focus on production costs. As these examples document

- a forecast presumes a purpose

Fundamentally, our purpose is not to predict the future precisely but to determine the effects projected results will have upon the scope, direction, profitability, and success of a given division, and of the corporation as a whole. Market conditions, however, steadily change, and therefore

- a forecast should not be rigid or inflexible

Its most significant contribution is to provide an identification of the *probable course* of coming events. In this probable course will rest the milestones characterizing planning decisions, including timing, commitments, costs, and expected results. Let's always keep in mind that forecasts are made for planning purposes.

I am often asked about the key to success in forecasting, and the answer is that it often lies on reasonable reliance on the simple laws governing an industrial enterprise and its environment. Worrying about complications by ruling out the possibility that the answer can be simple, would be foolish.

This simpler approach will typically be step-by-step, allowing better focus, in spite of *ifs* and *buts* that lead to deviations from the mainstream and (often) to alternative courses. Thinking by analogy, it is like studying an ancient coin.

Under ordinary light, chances are that features worn down by the coin's passage from hand to hand will be hard to make out. But by pointing a spotlight at it and illuminating the face of the coin from an acute angle, the resulting shadows emphasize some of the minor details.

This is indeed a technique helping archeologists reveal superficially invisible clues of any objects they uncover. The process is known as *polynomial texture mapping* and it provides an invaluable way to illuminate the past. The same concept is also instrumental in studying the future, at least up to a point.

## 4.2 ASSUMPTIONS MADE IN FORECASTS

The message Section 4.1 brought to the reader's attention is that archeology and forecasting share a common methodology, though archeology aims to uncover what has happened in the past, whereas forecasting's objective is to estimate (or guesstimate) what might happen in the future.

To serve in management decisions, forecasts have to be made with an acceptable degree of accuracy. This has much to do with the seriousness and documentation of our assumptions about oncoming events and changes—even if the latter are never completely correct. It is important to appreciate that every forecast involves tentative statements. Projections are not based on facts but on hypotheses.

- The more detailed these are, the more detailed the forecast,
- but at the same time, the less the likelihood is that the forecast is accurate.

These two bullets see to it that forecasts have limits, a fact to be fully accounted for by business executives who rely on them in making their decisions. On the other hand, although it might be sufficient for the manager of a small business to depend on its bosses' intuitive and unsystematic estimates, the larger company has a definite need for the methodical analysis of future possibilities.

Sometimes this analysis has bias, being unduly influenced by a trend or a popular issue. Electrical vehicles (EV) are a case in point. Their proponents say users will be happy with a restricted commuting range. To the contrary, a survey conducted by Deloitte Consulting in the United States found that 70 percent of potential buyers want a 500-kilometer capability.

Based on such studies and his own projections, in late 2010, Dieter Zetsche, Daimler's CEO, poured cold water on Nissan and Renault chief Carlos Ghosn's prediction that, by 2020, EVs would account for 10 percent of global auto sales. To Zetsche's opinion it would be more like one to five percent.*

The effective forecasting of future sales is largely a matter of determining factors that affect the sales volume of a product, as a function of direction and extent of major influences. When the direction is known, the amount of the product that is likely to be sold can be computed at a given level of confidence.

It is frequently, though not always, possible to predict how a small number of critical factors will change in the future; hence, the trend of sales. To do so, forecasters pay particular attention to prevailing interactions among

* www.automotivedesign.eu.com, November 2010.

the elements constituting the competitive future, exploring alternatives, and looking into contradictory as well as complementary solutions. A good guide is to compute a range of possibilities with their associated probabilities.

- Past statistics serve as a reference, as many events tend to repeat themselves over time.
- They also bring some objectivity into the forecasting process by reminding of events that took place.

By contrast, quite frequently forecasts based mainly or only on personal convictions reduce the associated amount of accuracy—all the way to predicting the opposite of what might happen.

"The biggest fool thing we have ever done...the bomb will never go off," said U.S. Admiral William Daniel Leahy, shortly before nuclear testing began in 1945. In 1932 American diplomat William C. Bullitt wrote to President-elect Franklin D. Roosevelt: "Hitler's influence is waning so fast that the government is no longer afraid of the growth of the Nazi movement."*

Maynard Keynes was an analytical minded, well-known economist, and active player in the stock market. Yet he failed to predict the crash of 1929 that led to the First Great Depression. "Who in the hell wants to hear actors talk?" has been the now famous statement made in 1927 by Harry Warner, founder of the Warner Bros. studios.

"I think there is a world market for about 50 computers," Thomas J. Watson Sr., founder and chairman of IBM is said to have stated in 1953 when he rejected the offer by the designers of Univac to buy their company. "There is no reason for any individual to have a computer in their home," answered Ken Olsen, founder and president of Digital Equipment Corporation (DEC), in 1987 when his assistants pressed the point that DEC needed to come up with a personal computer.

Biased hypotheses make forecasts ineffective or outright misleading. They could also lead to the company's demise, as it happened with DEC shortly after its years of glory in the 1980s as the No. 2 computer manufacturer in the world. The failure to forecast in a pragmatic, unbiased way has often led to corporate crises.

Companies that base decisions on fixed preconceived ideas or the intuition of the moment estimate quite wrongly the course of coming events.

* International Herald Tribune, November 30, 1988.

Precisely because of the need to instill confidence when confronted by uncertainty, the personality of he who is entrusted with the forecasting job can be of significant value (see also Chapter 6 on personality traits). The forecaster should possess

- an analytical, penetrating mind
- a background to appraise economic and business issues
- the experience necessary to make his rejections believable
- the ability to both quantify and qualify future potential
- skills in estimating trends, quantities, yields, costs, and profit potentials
- the ability to live with and even thrive with uncertainty, and
- the quality of making up his mind when faced with difficult, conflicting situations

Economists are famous for making reserves about alternative paths events may take. Anecdotal evidence suggests that President Truman said one day to his immediate assistants: "I am looking for a one-handed economist." When asked "why," he answered "So that he does not give me an advice and immediately add: 'On the other hand….'"

It is nevertheless a sound practice to develop alternative scenarios in forecasting—each with its own likelihood—when the situation warrants doing so. One way to develop alternative forecasts is to project the best, most likely, and worst conditions. Another way is to stick to the "most likely" as central value, and use confidence intervals to reflect the distribution due to variation of key factors.

Brief, focused comments improve a forecast's value to its user. If the prevailing conditions clearly show that a given market has escaped the company's hold, the forecaster needs to explain *why* this is the case. He has to be convincing in his arguments; for instance how fast the market is moving in a different direction.

A basic reason behind the failure to make pragmatic forecasts is a lack of methodology (Section 4.4)—all the way from making assumptions to producing computational results. Effectiveness in forecasting requires that deliverables are based on clearly stated assumptions that can be tested and retained, or discarded.

- Forecasters should never show top management what the latter want to see or hear even if, sometimes, top management pushes the forecast(s) in such a direction.

- Assumptions may be favorable or outright unfavorable to prevailing senior management's thinking, but either way the forecaster must be pragmatic and forthcoming.

Unrealistic assumptions are also being made when managers do not understand the value of working with alternatives, basing some or all of their decisions on confused estimates of the future or plain ignorance. When what management wants is "exact answers," the most likely result is that it gets patently false answers or no answer that can be of any service.

## 4.3 FORECASTS AND ACTION PLANS

Forecasting does not take place in the abstract. Effective forecasts look into the future impact of current decisions by paying attention to the company's strategy, its risk appetite, its planning needs, and special business characteristics. They also accommodate, reflect, and discuss contrarian opinions, and they express their findings in a comprehensive form.

The reference made to risk appetite may sound strange at this part of the book, but William Allen, Boeing's CEO, succeeded by taking risks. He also said that risks, which put the company in peril, are the most important job reserved to the chief executive. All other people in the organization—from vice presidents to middle managers had to work within established limits.

It is not generally appreciated that risk appetite can condition the breadth and depth of forecasts. We must reject a priori that something is impossible, Allen used to say, adding that integrity is the basic ingredient of authority. Within this perspective, forecasts are composed of three basic parts each connected to a specific theme:

- Economic mission
- Competitive strategy
- Action plan(s)

Relative to the *economic mission*, forecasts stress the kind of likely (but not sure) economic events that could affect the business, the firm, as well as its ability to perform. Forecasts aiming to help *competitive strategy* focus on the right product/market combination for effectively meeting objectives associated to business operations.

Forecasts made with the specific goal to assist an *action plan*, involve analysis of short-term developments affecting budgets, sales quotas, production schedules, supply chains, and other outstanding commitments. They aim to help management run its day-to-day business in a factual, comprehensive, continuous, and profitable manner—in knowledge of what, most likely, might lie ahead.

In every one of these themes, forecasters must present to management convincing and documented hypotheses. Therefore, of significant help to forecasting is a comprehensive information system endowed with corporate memory facility (CMF). The latter registers all assumptions, opinions, forecasts, and statements made by all parties to a decision, and it can be of invaluable assistance to postmortem walkthroughs as well as subsequent forecasts.

Postmortem walkthroughs are instrumental to the processes of focusing on an issue and pruning the forecasting methodology. An uninhibited two-way flow of information provided by means of interactive communications also establishes the necessary basis for *corrective feedback*. The latter is a vital ingredient of the ability of experienced managers to be in charge.

- The right feedback helps in redimensioning or altogether discarding overoptimistic and overpessimistic forecasts.
- In the longer run it can as well be instrumental in seeing to it that the assumptions being made become increasingly pragmatic.

A properly structured action-oriented forecast enables management to weigh potential benefits against potential risks; estimate the value to be (probably) derived from committing resources; and quantify the most likely outcome under different probabilities that important events happen. This is of great assistance to corporate strategic planning, and one of the most rewarding areas where forecasting can be of use.

To prove its value, an effective forecasting and planning procedure should allow for the possibility of errors in a forecast. The rule to remember is this: As long as management regards the lack of ability to "predict exactly" as an obstacle, rather than as an incentive, forecasting will be of limited value.

- The right forecasting mechanism implies a clearer understanding of a *likely* but *not certain* future impact on present decisions.

- Management has lots to gain by being aware of the likely future impact, because this conditions the direction and amount of commitments.
- But management is in the wrong track when it wants assurances and certainties about the future; by definition, *the future is uncertain.*

The forecast is only providing an opinion by trying to express such uncertainty in comprehensive terms. Subsequently, *plans* will characterize the interdependence of structures and resources pertaining to industrial activity.

As we will see in Sections 4.6 to 4.8, a formal plan leads toward the fulfillment of objectives, answering opportunities as they appear. An objective is an end; a result is the outcome of a specific task or a function to be performed. Objectives can be long range or short range.

- The long range both cover and help clarify those of a shorter range.
- Major goals determine the nature and timing of minor goals.
- Within the framework of management planning, they become statements of purpose evolving from the tentative expression of an action plan to its specific declaration.

This is possible because planning rests on a well-defined line of action with objectives to be met, resources to be committed in meeting these objectives, stages to go through, and methods to be used. While forecasting means assessing the future, planning is making provisions for it aiming to assure that established goals can be met

- within acceptable timeframes, and
- at an acceptable cost

In this sense, the plan is *directional*, providing the background capability for reaching chosen goals. A plan is efficient only if it brings about the accomplishment of goals with a minimum of unsought and/or unexpected consequences, and with results greater than cost.

Since forecasts change, plans must be *change oriented*. This helps to keep the plan up to date, and it also encourages a search for improvements. The reader should, however, notice that, in the general case, the notion of "improvements" is the *alter* ego of the notion of *change* (Chapter 3).

Another characteristic shared by forecasts and plans is that they must be open and *interactive*, assuring a steady interplay between all divisions and departments whose tasks may be different but they share a common goal. A forecast or plan held "close to the chest" will have little effect on the operations of an organization.

Last but not least, good forecasts and good plans are *time sensitive.* Scheduling key events is critical to orderly and efficient accomplishments. The right timing permits adequate consideration and coordination of the various steps involved, thus strengthening the focus and continuity of business operations.

## 4.4 FORECASTING METHODOLOGIES

The aim of this section is to explain what it takes to develop a valid forecasting methodology, not the "one and only" solution. There exists no unique forecasting methodology, but there are some principles that form a general frame of reference to be adapted to specific situations. As a starting point, a good forecasting methodology would involve three main components:

- Data collection and analysis aiming to identify trends, deviations, opportunities, and risks
- Field research—that is, estimates (for instance) of market potential, or of the market's feeling*
- Established indicators that may be leading or lagging, but still provide valuable information on a developing trend, or significant event likely to happen

Section 4.2 brought to the reader's attention that one should not believe, as some people do, that historical data is of little or no importance to forecasting. Quite to the contrary, it can make a great contribution, provided that we use statistically significant time series to simulate the real world situation. This approach is followed by tier-1 companies in areas such as

- economic projections
- policy decisions regarding investments, and
- monetary policy decisions

* Sometimes data analysis and the market's feeling point to the same direction. In other cases they contradict one another. Usually when this happens the market's feeling wins.

True enough, differences in methodology, model structure, and chosen variables often result in economic forecasts displaying significant differences. These may be due to the theoretical coherence of the model, estimation procedures, number and type of variables, or other factors.* Practically the same challenges exist with business modeling.

Precisely because such differences are current currency, central banks and other economic forecasting entities generally maintain suites of models, used in parallel for forecasting. The justification for that is that even sophisticated models are an oversimplification of the complex reality of the economy, and a battery of models can fill gaps left by one or two models. (At best, this is a doubtful statement.)

Modeling has become integral part of forecasting methodology because it helps in expanding the horizon of more classical empirical analyses conducted with a few factors. The theory is that even if the results being offered by successive simulations are not convergent, the forecasting effort can gain in analytical clarity as multiple modeling

- allows a better founded and more broadly supported judgment, and
- offers a polyvalent view of what might happen in economic or business activity, by capitalizing on margins of uncertainty

This is not always the case. What is true, however, is that simulators enable marketing personnel and production planners to experiment online, optimize the sales–inventory–production aggregate, examine necessary changes affecting the planned course of action, and analyze hypotheses that involve uncertainty as to the validity of suppositions underpinning them.

Experimentation provides a solid basis for planning sales, inventories, and production; the reader should nevertheless appreciate that a forecast is an estimate. By contrast, a plan is a commitment. The two may well complement one another (with the forecasting prerequisite), and they should be made with full knowledge that they should be kept dynamic. When this happens, forecasts are providing a basis for

- the eventual choice of a course of action
- better use of resources to meet commitments, and
- effective solutions to problems crucial to the company's future

* For instance, there may be no historical yardsticks for comparisons when assessing the underlying economic conditions, because rather uncommon singular events have been exerting a critical influence.

It is part of a valid forecasting methodology that its users appreciate that, though necessary, patterns of economic relationships derived from past events don't always form a sustainable basis for drawing conclusions about the future. For instance one of the assumptions which turned to dust with the 2007–2010 deep economic and banking crisis, was that developments in financial markets have no serious effects on cyclical developments in the economy.*

Still another consideration inherent to a sound forecasting methodology, is that model-based forecasts must be used with caution. This is particularly true of forecasts that rely on incorporating subjective empirical knowledge and/or unproved theories—like the one which maintained that markets are efficient.

Challenges connected to forecasting are less pronounced in their complexity when the subject is a specific sector of business such as marketing and sales. Sectorial domains do not feature extravagant theories. In addition, wrong premises or unwarranted assumptions are usually less controversial, though their presence cannot be totally excluded.

This fairly significant difference in objectivity between economic and business forecasts (sales, production, inventories) has to do with the fact that in the latter case the object of forecasting is much more contained. Typically, after a sales forecast is made, the natural progression is from "consolidated projected figures" to a

- sales plan,
- sales budget,
- manufacturing plan,
- manufacturing budget, and
- inventory policies

Supply chain commitments and unwarranted capital expenditures and/or operational costs can be avoided if management knows ahead of time the most probable course of market events that condition sales results. Apart from quantitative figures this requires a straightforward methodology as well as orderly thinking.

An industrial forecaster can start his work by first identifying business trends in the company's major product lines. He or she must then look at how

* Another example of a theory which has been shred to pieces by the same abysmal event is the "efficient market theory."

the firm has positioned itself in each market product-by-product, because inevitably its strengths vary by product and by market. Among crucial considerations to which the forecaster should be sensitive are as follows:

- Market potential, as it develops
- The competition's wares and their pricing
- Time-lag(s) in his firm's R&D deliverables
- Effect of time-lags on sales and on market share
- Investments and return-on-investment
- Coordination with other company plans affecting products, their markets, and their sales

There are no fortune-telling features in this process, in the sense of gazing through a crystal ball. The emphasis is on methodology, on data, on patient research. The strength of a forecast will be conditioned by how well the forecaster can predict "benefit" and "risk" from statistics, projections, and related research. Technology acts as an amplifier of the human skills, not as a substitute for them.

## 4.5 A FORECAST THAT WAS NOT HEEDED BUT THAT PROVED RIGHT

The fate of forecasts is far from being uniform. Some are up to the mark and contribute a great deal to effective planning. Others prognosticate the wrong path of events, or are used as an excuse why planning has little to do with reality. Still others have proved to be right, but the projections they made about likely aftermath were not heeded—and no corrective measures were taken in time.

This latter case has been the remit of a well-done study at MIT a little over two decades ago. Focusing on salient industrial and political issues confronting the American economy in the late 1980s, Dertouzos, Lester, and Solow stated that the stakes were nothing less than America's future wealth and political power; projected the reasons for this; and suggested some urgent changes.

The MIT researchers concentrated their study on eight industries, each of which got a chapter in their book.* They visited more than 200 companies

* M. Dertouzos, R. Lester, and R. Solow, *Made in America*, MIT Press, Cambridge, MA, 1989.

in the United States, Japan, and Europe and in their findings they detected a persistent pattern of U.S. management failures cutting across all industries: "American companies are no longer perceived as world leaders, even by American consumers," they wrote in their book.

To judge how much *on the mark* Dertouzos, Lester, and Solow were in 1989, the reader should know the opinion of an expert interviewed by Bloomberg News on April 6, 2010 on the occasion of the renewed Washington pressure on Beijing to revalue the yuan. "I don't think this would make much difference," the expert said. "In the preceding 5 years the yuan was somewhat revalued but the American deficit with China kept on increasing. Moreover, much of what we import from China we do not manufacture anymore."

Many of the American goods-producing companies that the MIT study said were endangered have, by now, gone out of business. This has been precisely the alarm bell rung by the researchers 22 years ago who also added that fixing the ongoing decline of American industry will take "wrenching changes at all levels of organization." It is not only America that finds itself in such straits.

The whole West is going downhill because of the same reasons that find their roots in past and current policies—made by second raters who pose as chiefs of state but in reality have been systematically destroying their country's industrial base. Businessmen with practical experience do appreciate the need to safeguard the future potential of industry, but politicians in power don't. Their forecasts are near-sighted; therefore they engage in

- unaffordable superleverage, and
- unsustainable mountains of debt accumulated through ever-growing budget deficits and current account deficits

What Dertouzos, Lester, and Solow prognosticated for America in 1989 was already in a state of drift in Greece, leading to its near-bankruptcy in 2010. For all Western countries this should be 12th-hour warning. Small economies crash faster because their resources get depleted easily, but big economies also reach the edge of the precipice if they continue in the same rotten path. Look at the 2010 Britain.

The lesson to be learned from the aftereffects of the unheeded forecast discussed in this section, is that *if* we are not able to reverse current trends and safeguard our future, *then* both ourselves and the next generation will pay dearly for it—whether we live in a small country or in a big one. With trillions of dollars spent to save big banks from self-destruction, given the state of indebtedness they brought to themselves *if* the United States, Britain, or Japan were business enterprises, *then* the only solution would have been filing for protection from creditors.

Back in 1989 the MIT researchers did not have the evidence of this unmitigated financial catastrophe—but they had the hindsight because they had experienced the 1987 stock market crisis and saw the trend. What we know now validates the researchers' forecasts. Western governments should get busy in safeguarding at what still remains in resources by

- reorganizing
- downsizing
- deleveraging, and
- punishing the wrongdoers to give at least a sense of justice and an evidence of ethics

One could echo as an excuse for inaction the eighteenth century wit: "Why should we legislate for posteriority? What has posteriority ever done for us? But is it sound to pose such a question?" As businessmen we should appreciate that deep moral issues are involved. And because we belong to human communities of one sort or another we come to share common obligations arising from the interdependences they involve.

This helps in reflecting on our obligations to future generations which, to a significant measure depend for their survival on our current actions (a statement also true in ecological terms). Our obligation to the future can even be justified in terms of self-interest. Contrarians might respond that there are deep reasons within our existing culture for thinking that it is very difficult to provide a moral basis for an *obligation to the future.* But this is big-headed. No generation should leave the economy and planet Earth worse than it found it.

## 4.6 PLANNING

The *planning* methodology targets the matching of resources—human, financial, material (factories, machinery, raw materials, finished goods)—against specific objectives over a given timeframe. Plans must be well documented but also flexible and adaptable, but most important of all is the methodology. "The plan is nothing," President Eisenhower once said, "planning is everything."

- Like strategies, a plan should fit the objective(s) of the people, enterprise, or society developing.
- By contrast, forecasts must be made on a given theme, but they should not fit any objectives; the measure of their quality is objectivity.

Well-managed companies place emphasis on *planning as a system*. Since an organization's present strengths often lie in its geographic or product leadership characteristics (or alternatively on its diversity) corporate planning should integrate among its "goals" coordination and teamwork aimed to preserve these assets.*

- The planning methodology is more important than the individual plan.
- Plans can and do change; the methodology, if correct, should only evolve slowly.

The more globalized the economy and the more competitive the market, the more planning becomes one of management's basic functions. Properly executed, it assists in seeing a clearer picture of resources and commitments, avoiding decisions that tie one's hands to a committed course of action over too long a period of time. Planning also promotes business opportunity and keeps risks under watch by

- forcing a consideration of alternatives, and
- obliging the organization to account for likely developments (see the discussion on forecasting)

Alternative plans, which are integral part of a sound planning methodology, contribute to better adaptation by highlighting the different ways available to accomplish an objective. Management has plenty of

* This does not contradict the need that the firm is able to reinvent itself, for the simple reason that the purpose of reinventing is the safeguarding of assets.

opportunity to study and choose the best course, because a plan is made in stages. There are nine main stages in corporate planning:

1. *Translating the goals established by board of directors, chief executive, or executive committee, into action plans*

Within the realm of overall company objectives, goals need to be detailed and assigned to functional bodies such as divisional authorities. Strategic objectives are set at the top management level of the firm; tactical goals, subordinate to overall corporate goals, are set at divisional level. All actions pertaining to both of them must be planned.

2. *Providing a framework enabling future decisions to be made rapidly and economically, and with little disruption of the chosen business course*

It is not enough to plan once and forget about it; no plan is for ever. Constant vigilance is the price of success, and vigilance involves repetitive opportunities for making decisions—whether within the framework of the original plan or, as needs and requirements develop, by altering it.

3. *Making choices among alternatives in order to achieve company goals, taking due account of the company's resources and programming their application*

Any company, like any individual, community or state, has finite resources. Unlike what some people tend to think, the most scarce resource in an organization is not finance but talent, expertise, and skill; all of which must be properly allocated to worthy goals.

4. *Assisting the staff in detailing chosen activities ranging from R&D to manufacturing, sales, and finance—both domestic and international*

Many of the actions promoted and supported by a plan will be decided by top management only in general terms. Middle management will bring to the plan detail. Because the devil is in the detail, it is advantageous to develop "Strengths and Weaknesses" reports for each market and for each product.

5. *Foretelling possible handicaps in each course of action, thus enabling top management to make decisions having in mind not only future opportunities but also risks and limits*

Since any forecast is based, at least partly, on tentative statements and hypotheses that find their way into the plan, the latter should be thoroughly tested with the aim to prove its worthiness (and unearth its) handicaps. The aim should be to prevent the usual "fire brigade" approach when adversity strikes.

6. *Setting the framework within which management will be able to make the maximum use of organizational resources*

A plan will be more effective if it provides management with clearly defined alternatives and choices falling within its framework. This way, after the higher-up choices are made, middle management can use its options to allocate and exploit the resources under its control.

7. *Evaluating the future implications of the established plan, prior to authorizations, and define the kind of decisions most likely required down the line*

The time to test the "what if" questions is during the formulation of the plan. Any proper plan is results oriented, but the search for concrete deliverables often brings along certain implications. At some point in time, negative aspects might outweigh the benefits to be created by achieving a certain goal; management action might upset existing balances; or unprecedented handicaps may require shifting resources from other plans (and objectives).

8. *Evaluating the results being obtained as the plan goes into action through controlled feedback, including individual performance appraisal*

Reporting on the plans' execution should be timely and structured in a way to bring to light the existence of deviations and exceptions. A monolithic plan masks deviations and is opaque in regard to general progress. The opposite also is true. Corrective action is blurred by too many requests for attention, because of undue detail.

9. *Timing all actions (and their phases) involved in a plan helps in identifying the multiple dimension of the planning premises and need for smooth transition from one stage to the next*

A basic requirement of every planning effort is to take account of lead time. Things don't happen instantly because they are part of a plan. Also in the long term the forecast(s) on which a plan is based may change. Therefore, planning must be flexible and readjustable in its functions to provide a basis for taking action, and as much as possible minimize the unknown. Several reasons may see to it that a plan fails. The most important are as follows:

- Lack of skill

This has many aspects. One of them is absence of a sound methodology, often coupled with substandard personal discipline—which is cornerstone in planning. Another is inability to focus and explore opportunities, or in spelling out risks and limits.

- Lack of market intelligence

As the careful reader will recall from Chapter 1, strategy is a master plan against an opponent. In business, these opponents are product and market competitors who get ahead of the curve. As far as planning is concerned, market intelligence is not an option—it's a "must."

- Absence of clear objective(s)

A plan characterized by unclear objectives, or one which tries to hit several goals at the same time, is doomed to failure. Effective plans have locality; they are specific and therefore they cannot be all things to all people (which is a basic reason why so often government plans go against the wall).

- Low technology

High technology, low cost, and rapid innovation are the three most important advantages in modern industry. In my professional practice I have often come across ambitious plans by companies whose information technology is no match to that of their competitors—and therefore unable to provide the instruments needed for planning, execution, and control.

## 4.7 PLANNING PERIODS

Top management needs *a view of the future* regarding products, markets, and actions needed to reach established goals. As we have seen in this chapter, forecasting projects on future events and planning helps in programming the management effort within a specific timeframe. Theoretically, timeframes may have any one of various ranges; practically the following four are the most frequently used

- *short range.* Twelve to 18 months, often called the "rolling year." The 12-month financial plan is the classical budget.
- *medium range.* Usually two to four years, generally not exceeding this period.
- *long range.* Five to seven years, varying with the enterprise, sometimes extended to 10 years.
- *far out* or *very long range.* The 15 years plan is the most recent, aiming at answering the questions, "What should this company be doing at 2025? From where will come the earnings?"

In the background of management planning in every one of the four ranges being discussed, lies the recognition of possible implication of commitments being made. When firm commitments are established, the logical course of action is to attach them to their chosen goals, with subgoals to serve as stepping stones.

Variations in the aforementioned ranges can and do exist. As we will see in the following paragraphs, the specific length of the planning period for developing formal, comprehensive plans is relatively unimportant. What *is* important is that this period—once established—is observed throughout the organization so that the time element is not disruptive.

Strategic plans are longer range. Medium-range plans typically address markets and products, and as Figure 4.1 shows can be strategic or tactical. Purely tactical plans are shorter range, mainly operational or financial.

**FIGURE 4.1**
The plan is nothing. Planning is everything, and ranges from a structured to an unstructured environment.

No matter which its range may be, a plan forces managers at all levels to be consciously alert to overall objectives and policies. This is particularly important in large organizations where complexity and lack of global standards are negative forces.

Programming for the far-out planning period is a most challenging enterprise requiring clear concept and vision. A 15-year range can be approached mainly from the viewpoint of general direction based on critical order of magnitude evaluations. Even so, however, it permits positioning our firm (in terms of future products and services) against market forces.

Many people object to far-out planning by saying that customer needs and requirements cannot be projected so much ahead of time. That's the wrong argument. In the early 1970s GE's far out planning unit in Santa Barbara, manned by GE's former general managers, had projected that by the mid-1980s more than half of GE products will use a microprocessor. It happened exactly like that.

Although the uncertainties increase as the time span is extended, the possibilities of considering various alternatives and maximizing choices is also greater. Emphasis is on adjusting to the future rather than simply assuming that present conditions will continue unchallenged and unchanged.

Probably the most difficult, but also most interesting and rewarding, aspect of far-out and long-term planning is the development of a program that shows how and under which conditions overall organizational objectives can be accomplished. The essential characteristics of such a master plan are that it is

- comprehensive, covering all major elements, and
- integrated into a balanced and synchronized program.

Another reason why long-range planning is one of the most challenging time spans is because it involves so many uncertainties—whereas problems in shorter-range planning can be traced to the absence of a clear sense of direction or of well-established priorities. Priorities are the substance of a plan, though they may also be hierarchical symbols.

By contrast, to what the preceding paragraphs brought to the reader's attention, the short-range plan is made after a "freeze" is taken on the consideration of possible alternative courses of action outlined by the medium-range plan. The focus of the short-range plan is implementation; whereas the medium-range plan's aim is weeding out a plethora of possibilities that may be long on promises and short on feasible, tangible results.

There is as well another valuable purpose for using medium-range planning: that of establishing interim objectives between long-term goals and those used in the development of programs for annual operations and budgets. *Targets* are specific results associated to medium-range milestones.

Planners should as well be aware of the fact that, although necessary, focusing on targets alone makes an incomplete plan. Selecting a course of action will not guarantee all by itself correct timing (Section 4.6). Milestones are required and their definition may be divided into three types:

- Those dealing with repetitive actions
- Those targeting a *one-tantum* event, and
- Those relative to a changing set of circumstances

Timing the actions of managers and professionals is relatively easy when these are repetitive and routine, and when a clear-cut procedure has been worked out for handling them. One-tantum events may also be crisp. But it is a challenge when actions and decisions are novel and unstructured. The latter tend to increase in number and in inherent uncertainty, particularly when the time span over which their impact may be felt is long.

This is one of the fundamental reasons why the span of detailed planning is limited to the period of 1 year, or that of a rolling year, in which predictions bear a reasonable degree of validity. However, since the certainty required in management planning varies with the type of industry or of a program, the choice of level of detail in each time range cannot be stated in a generally accurate way—except for the fact that short-range plans must always be both detailed and very well documented.

## 4.8 INTEGRATIVE PLANNING: A PRACTICAL EXAMPLE

The principal aim of instituting a planning function is the orderly execution of business activities, matching resources to tasks and timetables. The plan provides a quantitative roadmap for management action, allocation of corporate assets to goals to be achieved, and their orderly distribution over time. The result is improvement of performance throughout the organization by means of

- operational coordination, and
- feedback permitting to compare deliverables to the plan

The effectiveness of management improves most significantly through the establishment and implementation of corporate and divisional programming procedures. The measurement of deliverables against target keeps people on the run. Any plan, be it long, medium, or short range will feature as focal points those pertaining to the division which establishes it.

- In *research and development*: research projects, breadboard models, development projects in the pipeline, R&D support to current product lines
- In *finance*: budgets, cash flow projections, profit margins, return on investment, return on equity, return on assets, admissible debt levels (gearing), capitalization
- In *manufacturing*: production plans, schedules, standard costs, cost reduction, just-in-time inventories, degree of equipment utilization, overall efficiency improvements
- In *marketing*: area of operations, market leadership, share of the market, effective customer handling, sales quotas, annual growth in sales, profit margin by product and product line in sales operations

The corporate plan itself must be *multidisciplinary* encompassing all company functions: R&D, manufacturing, marketing, finance, and human resources. Because the organization of the typical company is departmental, it is essential that the planning mechanism provides for sectionalization as well as for subsequent coordination between departments.

The planning framework can best match the firm's specific divisional structure if it is guided by properly defined objectives. In Peter Drucker's words, "Objectives are needed in every way where performance and results directly affect the survival and prosperity of the business. The imagination and depth at which this is done will largely determine the quality of the whole management planning and control process."

Lack of appropriate definitions leads to increased frustration about how to best take advantage of available resources and market opportunities. It also makes more complex the job of integrating the plans of the various departments, divisions, and regions in a way that assures management commitment and involvement at all levels. This integrative planning process comprises four phases:

- Building the information base that registers all divisional contributions
- Assuring that the objectives for the year (or for the medium term) are fully understood by all concerned
- Integrating the operational plans of individual operating units into a well-coordinated aggregate
- Assisting the management control function of the organization (Chapter 7) in monitoring and evaluating performance against the plan

As far as integrative planning is concerned, some companies choose centralization whereas others prefer decentralization. My approach merges these two alternatives making the development of operational plans for individual units the responsibility of each manager in charge of an operating unit, but giving to central planning the responsibility for integration and coordination of individual plans.

For each unit, the end result must be an operational plan that comprises analysis and conclusions relating to local conditions, specific action steps to be taken, and targets that the unit plans to achieve within the firm's defined mission. Planners should appreciate that there exist as well constraints in aggregating a corporate plan based on individual operating unit plans.

Here is a practical example on how a medium size organization develops and implements an integrative planning effort. Divisional contributions for each planning cycle begin about 6 weeks before the plans are due at corporate headquarters for consolidation. The planning process starts at divisional level—or, more precisely at the profit centers. There are 100 profit centers in this firm, the CEO's credo being that nothing stimulates more than profit management. This

- gives the responsible executives the full flavor of the business, he or she is in charge, and
- provides each executive with the opportunity to identify himself with the policies and philosophy of the company in a demonstrable way

At each division, the development of its plans begins with sales forecasts; an area that has been perfectly mastered by this firm. Each division then works backward from sales projections to production schedules, manpower requirements, capex, procurement, and budgets.

Forecasted sales factors to be used for inventory and production planning are broken into several elements. Whenever possible, it is almost on a salesman-by-salesman basis. Once sales managers compile the input of individual salesmen, reviewed and corrected by the local sales director, division managers combine this data with market research results. Subsequently, production planning sets production capacity needs. If an expansion or cutback of capacity is judged necessary, this is noted.

Sales forecasts are also employed to predict when changes in sales targets or quotas may occur. This input permits to fine tune cash flow projections. In the more stable products, the cash flow picture is detailed 3 years ahead. In developing their plans, division heads estimate likely variations and cite reasons for them. Among planning factors are salaries, materials, markets, transportation, overhead (that is always kept low), and any unusual situations such as strikes or adverse weather.

Division heads discuss their prospective plans with group executives, and coordinate with related divisions. This leaves room for few surprises since managers are in constant contact. Plans flow upward to corporate headquarters where a financial analysis and review team bundles them into a total corporate package.

- This becomes the main reporting vehicle to the CEO and board of directors.
- The plan and control feedback are the means by which the board controls operations.

The integrative plan and its divisional details are presented to the board as recommendations. In cases of major decisions, such as capital spending in excess of target limits, the report would contain both the reasoning behind the recommendation and alternatives. Profit plan goals and past performance are listed in terms of profit centers: First, the company as a whole, then by groups, and then by divisions. Groups presidents, who are all board members, are on hand for any explanations.

This procedure which, in the firm in reference, has become an established methodology, is followed by critical evaluations of the profit-making potential within each area of operations. Qualitative criteria, like a "cash cow" profit center, bring into the process an important element of performance. Just as important is the quantitative evaluation.

Different measurements have been instituted, to permit objective measurement of marketing effectiveness. Their aim is to identify attained profitability measured against profit-making potential, steadily checking the cost of goods sold. At times this feedback is critical, but it is also strengthening the distributor organization's ability to improve its deliverables.

# 5

# *Organization and Structure*

## 5.1 STRUCTURE MUST FOLLOW STRATEGY

The future belongs to those who can create the structures needed for today's and tomorrow's more complex and more uncertain business environment. When in the middle of the first decade of the twenty-first century banking and financial services became a battleground, structures unfit for that environment crumbled taking the institutions down to the abyss. There is no better example than that of big institutions: Citigroup, AIG, Bank of America, Royal Bank of Scotland, UBS, which were characterized by non-integrated units resulting from mergers and acquisitions with diverse

- cultures,
- strategies, and
- structures

The man who, back in the early 1960s, first linked structure to strategy was Alfred Chandler.* When he died four and a half decades later, in May 2007, he was acknowledged to be the dean of American business historians—the man who more or less invented the history of the big corporation and derived lessons from it.

In his seminal book *Strategy and Structure*† Chandler underlined the concept that structure must follow strategy, and structuring a company should be done in a way providing a sharp practical edge. His thesis was as well that changes of strategy can be successful only if managers are willing to cast their organizations into new forms.

* For much of his life, Alfred Chandler taught at Harvard Business School, where he made business history mainstream.

† Alfred Chandler, *Strategy and Structure*, MIT Press, Boston, 1962.

To Chandler's thinking the big and powerful corporations arose in the late nineteenth century as a result of the integration of what started to be mass production and mass distribution. Their ascent was based not only on financial power but also, if not mainly, on superior management techniques—whose beginning can be traced in that same period.

Indeed, the late nineteenth century was prolific in more than one way. The first concise theory of management is credited to Henri Fayol, the CEO of a major French manufacturing enterprise. The first organized industrial research laboratory was made by Werner von Siemens,* the founder of the synonymous German industrial empire.

The same is true of the early twentieth century. The original concept, and first application, of time study aimed to measure performance and increase worker productivity was made by Frederic Winslow Taylor, an engineer at U.S. Bethlehem Steel in 1912. The first motion study is credited to two American industrial engineers, Frank B. and Lillian M. Gilbreth, and dates back to the mid-1920s.

It is by no means a coincidence that also in the mid-1920s Alfred Sloan, the legendary CEO animator and chairman of GM, invented the framework as well as the principle of line and staff, which today characterizes the structuring of the corporation. Shown in general lines in Figure 5.1 is Sloan's solution, which consists of a

- headquarters staff that plans and coordinates and as well provides R&D, engineering, marketing, and financial services to the divisions, and
- a number of independent divisions whose main objective is to uphold their distinct trademark and channel the motor vehicles to the market.

The divisions feature smaller staffs operating under their control of the above mentioned central services. One of the tasks to which Alfred Sloan put the concept of divisional structuring was to provide a thin red line dividing one product line from the next. Originally, the motor vehicles

---

* Werner von Siemens (1816–1892) established his first company in electrotechnics in 1841, moved into telegraphy in 1846 and in 1847 made (with Johann Georg Halske) the *Telegraphen-Bauanstalt Siemens & Halske*.

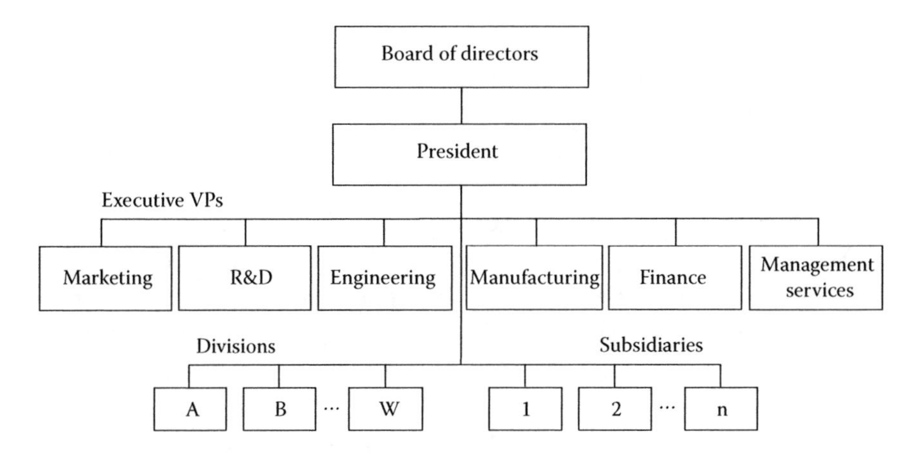


**FIGURE 5.1**
The concept of corporate organization and structure dates back to the 1920s, and can be credited to Alfred Sloan.

of companies aggregated into GM competed directly with Ford's Model T and therefore appealed more or less to the same customers class. This was counterproductive.

Sloan's penetrating mind understood that such overlaps did not make sense. Each motor vehicle trademark should have a well-defined market to which to appeal. Cadillac was assigned that of high net worth individuals. Next to it came Buick followed by Oldsmobile. The overlap in market segment and appeal was minor.

Oldsmobile addressed the upper mid-market for autos; Pontiac was invented to cover the lower mid-market. The wider population of current and potential motor vehicle owners found at the base of the market was the remit of Chevrolet—which, for any practical purpose, directly competed with Ford's Model T. As this example demonstrates

- a properly thought-out organizational structure helps, through target products, defining market segmentation, and
- among them these products and market segments give substance to the concept of the corporation as a productive, wealth-making entity; and also serve its strategic goals

It is quite likely that Alfred Chandler had been influenced by Alfred Sloan's emphasis on corporate structure, since he collaborated with him in the writing of Sloan's excellent book *My Years with General Motors.** It is also probable that the need for structure came to him as the result of observing the complexity of relentlessly internalizing a giant company's transactions, which had previously been done piecemeal by separate, smaller independent firms.

Confronted with these challenges, the mind of an alert person evidently appreciates the need for developing and adapting new organizational forms to deal with problems that continue to evolve and expand. The need for structure became clear as multidivision firms welded separate business units into an integrated and coordinated enterprise with unity of command.

On the basis of the evidence that he collected, Chandler argued that from the late nineteenth century onwards the role of coordinating economic activities passed from proprietors to the visible hand of professional managers. They built, led, and ran the big organizations characterized by a fairly disperse ownership—the shareholders. To this it should be added that the industries that drove economic growth for much of the twentieth century were dominated by a small number of vertically integrated firms modeled after the German Konzern. Capitalizing on

- their size, and
- the financial resources they brought under their wings

these large-scale corporations continued to grow and master the market, steered by professional managers. Indeed, in contrast to the late nineteenth century, which was the age of proprietorship, the twentieth has seen the rise of *managerial capitalism*. Many people now believe that this epoch has come to an end.

- Successful companies are getting smaller while giant firms drive themselves against the wall.
- Proprietors start rising as a managerial class, taking over from the class of "managers for hire."

The archetype of dinosaurs, as the 2007–2010 deep economic and banking crisis revealed, has been the big banks "too big to fail." Their descent to the abyss documented that "too big to fail" also means too big to be

* Alfred Sloan, *My Years with General Motors*, Pan Books, London, 1969.

managed. As for the professional managers they have stepped on banana skins of their own making: higher and higher salaries, more and more bonuses but less and less professional skills.

This is indeed a big switch from Alfred Chandler's thesis on corporate structure, but it hardly makes his views irrelevant. There is no better way to understand the era of organizational unbundling than to study the era of superbundling that came before it, including

- its successes,
- its failures, and
- its shortcomings.

By all likelihood, during the coming year the organization and structure of companies will change in form. The careful reader will recall Drucker's thesis about a new structure resembling, most likely, that of an orchestra with one conductor and hundreds of professionals. But there will always be a structure.

## 5.2 RADIAL, SPOKELIKE ORGANIZATIONS

In putting together in the second half of the nineteenth century the first major multinational corporation, Standard Oil, John D. Rockefeller created a forerunner of the megafirm. At the top of its might, before antitrust broke it up, Standard Oil refined, distributed, and marketed nearly 90 percent of U.S. oil and it provided a model of what later became known as radial, spokelike organizations.

Rockefeller invested in his first refinery during the American Civil War. When he cofounded Standard in 1870, a chaotic pricing pattern prevailed in the oil fields of western Pennsylvania, then the heart of this industry, with gluts dragging down prices below production costs. Rockefeller came around that mess through a strategy that made him a legend—he

- bought up rivals,
- modernized plants,
- imposed a discipline on the industry, and
- organized the oil business on an enduring basis.

This strategy required supreme cunning and formidable self-control, first of all by John D. Rockefeller himself, who was still in his 30s—a young wonder of a new business being born.* Apart from discipline, it also required a framework allowing to keep the sprawling company in one piece and counteract the centrifugal forces that were always present. This was provided by John Rockefeller, Jr. who succeeded his father at the helm of Standard Oil.

- The father was a visionary, the son an organizer, and
- as an organizer, he preceded by a short period of time Alfred Sloan (Section 5.1).

One of the characteristics of both Rockefeller and Sloan was allegedly their aversion for traditional bureaucracy with defensive positions and inertia. This is not surprising because empire builders are unanimous in their appreciation that bureaucrats thrive by routinizing tasks, and routinized tasks, which they repeat over and over, identify companies that are not likely to grow and thrive.

This reference to the psychology of industrial tycoons is important in as much as the choice of great entrepreneurship is to that of a dynamic structure. They place themselves at the top of the pyramid, or more precisely at the center of the web, with the chain of command radiating out spokelike in all directions. This organizational style

- keeps the entrepreneur at the heart of action
- permits easy communication and networking, and
- encourages innovation by short circuiting the hierarchical structure

Great entrepreneurs favored the web because it permits to run the system fast and effectively. One of the flat organization's great advantages is rapid adaptation to market changes. As Sam Walton, of Wal-Mart fame, used to say, a dynamic organization must be able to turn on a dime.

By contrast, long hierarchies have so much inertia (let alone unnecessary costs, Section 5.3) that it takes a long time for the system to turn around. This significantly delays, and quite often blurs, the needed response to

---

* When Rockefeller's net worth peaked at $900 million in 1913, it was equivalent to more than 2 percent of the gross national product of the United States. No other person before or since made that feat.

market changes. It also makes it more difficult to clearly identify realistic goals for each business unit.

The fact of life however is that after the empire builder is gone hierarchical power, and with it bureaucracy, takes the upper ground, and the two-dimensional web becomes a string of cogs. It has happened for example in GM, after Sloan was no more in command. After Watson Sr. IBM followed the same path, and this is true of practically all other corporations.

Over the last 100 years, three types of industrial and financial corporations have evolved from the original lean staff and line structure of Sloan and Rockefeller—which, as Figure 5.2 shows, was two-dimensional, as practically all entrepreneurial radial, spokelike structures.

- The three-dimensional structure of the 1960s added a higher-up layer of bureaucracy through the holding.
- The four-dimensional structure of the 1980s expanded to benefit from globalization.
- The federated style of the 1990s has followed an original Japanese concept in an effort to find an organizational solution for virtual corporations.

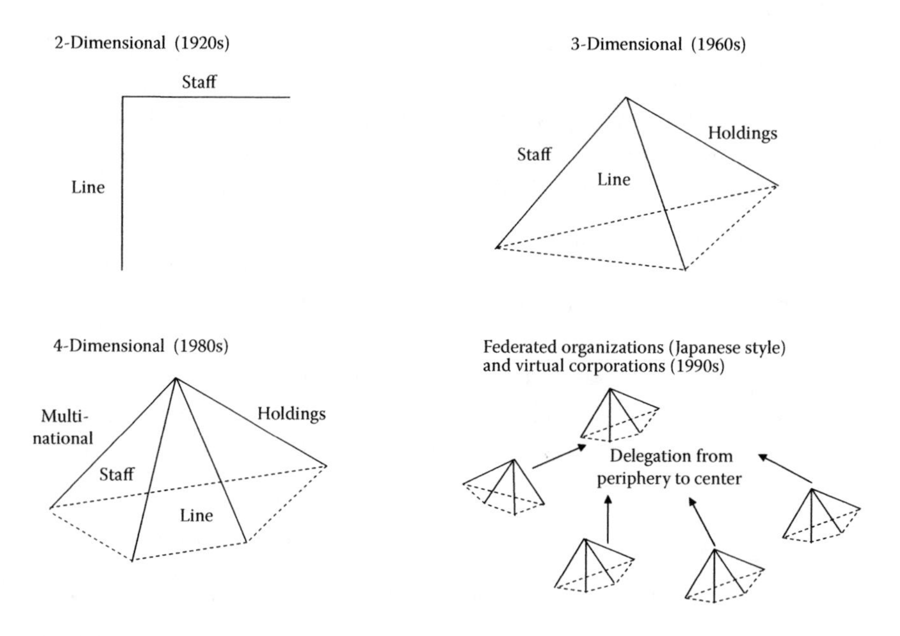


**FIGURE 5.2**
Eighty years of evolution in the dimensions of industrial organization and structure.

Responsible for the three-dimensional model has been the wave of mergers and acquisitions in the 1960s, and popularization of holdings (often promoted as a way to bypass different legal and other constraints put in place to control big size). One more dimension has been added in the 1980s with the aggressive multinational expansion of big firms and the advent of globalization.* (More on structures needed for globalization in Section 5.4.)

Globalization, of course, has been an invention of the nineteenth century, not the twentieth. But the early internationalization of business was largely a merchandizing affair characterized by exports (from the industrial countries) and imports (mainly mining). Companies started seriously considering, and also building, manufacturing operations abroad prior to World War I. With few exceptions, however, this policy was put on hold during World War I, the intervening war period and World War II.

Globalization became a strategic solution in the 1980s. Holdings and globalization significantly complicated the corporate structure. They also reduced top management's hold on the corporation (despite claims to the contrary), and made information on the web of operations too opaque.

Opaqueness has been also promoted by complex financial instruments. When on April 7, 2010 Chuck Prince, the former CEO of Citigroup, and Robert Rubin, former Treasury Secretary and former chairman of Citigroup's executive office, were summoned to testify to the U.S. Congress about Citigroup's travails, they said that they did not know about their bank's huge exposure to collateralized debt obligations† (CDOs) till its descent to the abyss.‡

This statement by Citigroup's former top brass also confirmed that, as it has been already brought to the reader's attention,

- companies too big to fail are too big to be managed, and
- the more complex are the instruments in which they deal, the more mismanagement takes over till the firm goes bankrupt.

---

* D. N. Chorafas, *Globalization's Limits. Conflicting National Interests in Trade and Finance*, Gower, London, 2009.

† Collateralized debt obligations. D. N. Chorafas, "*Financial Boom and Gloom. The Credit and Banking Crisis of 2007–2009 and Beyond*," Palgrave/Macmillan, London, 2009.

‡ Bloomberg News, April 8, 2010. *If* one can believe that the bank's two most senior executives did not know about CDO risk, *then* he can believe anything.

CEOs who think that they can thrive on complexity have been wrong 10 times out of 10. The four-dimensional enterprise is not a novelty or a beauty; it's a curse. The "right size" is one so aptly defined by Sam Walton when he said that management must be able to put its hands around the problem.*

Figure 5.2 also presents to the reader a four type of structure; the federated organization. Its special characteristic is that authority does not flow from the top to the bottom. Instead it is delegated from the periphery to the center, allowing it to act as a "conditional top."

In its original form, this structure has been, and continues to be, used by Japanese *Keiretsui* (conglomerate), but corresponds to a solution that is alien to the West. It has been tried, for instance, with the European Union (EU), in the hope that it will lead to better integration. Instead the result has been disintegration, augmented by the unstoppable enlargement of the EU. Nobody seems to have explained to the 27 chiefs of state that growth for growth's sake is the philosophy of the cancer cell.

## 5.3 THE SPAN OF MANAGEMENT

The only true measure of the size of a company is turnover. The number of people working for it does not mean much, because inefficiency may be rampant, with the result that many employees are loafing. Profit is no measure of size, either, and neither is capitalization.

Modern companies such as Google and Amazon.com are built on turnover. Another important fact, of which to take note, is that they are not sprawling organizations where top management loses contact with the base and with the market. In a structured sense they are flat, radical organizations with very few middle management layers.

Contrary to the remnants of nineteenth and early twentieth centuries' big brown industries (steel, autos, mechanical engineering, chemicals), modern companies are hard-pressed to find the most efficient guide for forming what many consider an essential building block of an adaptable and profitable structure:

- The smaller highly competitive companies know that keeping dynamic is vital to their survival.

* Sam Walton, *Made in America: My Story*, Bantam, New York, 1993.

- Among the bigger firms, the wiser ones appreciate that the ability to reinvent and renew themselves is instrumental in facing problems born of market gyrations and complexity.

As the second decade of this century unfolds it will become increasingly evident that every company must be able to revamp and remake itself to survive in the unfamiliar world of saturated markets and borderless competition. The cases where management is standing at the crossroads between the past and the future multiply. Hence daring is required and this calls for a slick management structure.

Peter Löscher describes as follows his action at Siemens, after he was named CEO in 2007: "One of my first sentences in the company was speed, speed, speed." Not everybody could follow. Half of the old operational management was moved out. The company's eleven divisions were reduced to three; and in place of management by committee and consensus, Löscher gave each of his *sector* (restructured major divisions) executives full responsibility for their businesses telling them:

- "It is about aligning accountability and responsibility.
- "At the end of the day I look into your eyes and say: you are responsible."*

Battleships cannot easily make U-turns, and the more management layers are featured in a structure, the more it resembles an old immovable battleship. In addition, the higher are the costs and the greater is the sense of false security felt by management and its underlings. This becomes highly detrimental because management's ability to run the organization is based on four spans:

- Span of control
- Span of attention
- Span of knowledge
- Span of support

Summing them up, they make the *span of management*. *Span of control* and management efficiency correlate. Simply stated the span of control is measured by how many people report to a higher-up boss. Two people, for example, is awful. It represents a highly pyramidal organization like the

* *The Economist*, September 11, 2010.

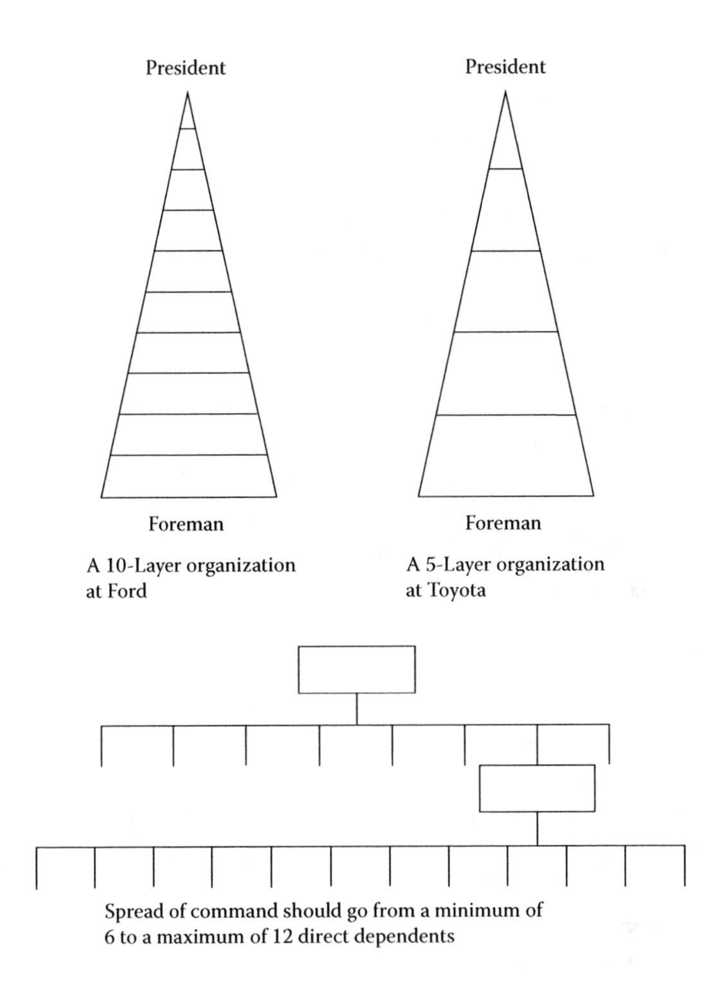


**FIGURE 5.3**
A lean organization has broad span of control and few middle management layers.

one in the upper left of Figure 5.3, featuring a great deal of intermediate layers from president to foremen.

Many years ago I was in a meeting with a senior executive of Ford who said that his company had a 10-layer organization. This, he added, was an improvement from 12 layers earlier on—but still too bad compared to Toyota whose management featured five layers. In this meeting, the Ford executive characterized Toyota as his company's major competitor, and added that Ford had to become just as lean.

In old times a narrowband span of control was more or less justified in order to permit enough management attention paid to each of the executives reporting to the boss. This is no more the case today as expert systems, agents, and real-time information technology are able to support and expand span of control, and therefore substitute several middle management layers. Few people really appreciate the many advantages of a lean organization.

As Table 5.1 shows, a company of say 30,000 people will make do with five layers with a span of control of eight, but it will need six layers with a span of seven, and seven layers with a span of five. Apart from the greater flexibility and lower costs that it provides, a greater span has important aftermath in making the organization more lean as well as better able to respond to changing situations that require repositioning against market forces.

The *span of knowledge* can be improved in two ways. The one is the proper intake of skilled, well-trained people and the provision of facilities for lifelong learning (Chapter 6). The other has to do with the proper setting of structural issues—most particularly with delineating effective relationships and establishing positional qualifications. This is achieved through job descriptions that emphasize

- responsibilities
- indicators, and
- objectives

The *objectives* are short term; they span over 1 year as already noted on a previous example with General Electric Bull. *Indicators* are classifying the nature of responsibilities, defining liaison lines between managerial positions in order to facilitate a better comprehension of complementary coordination.

*Responsibilities* should be defined in a few bullets, say 2 to 4; not just 1 but neither 5 or more. In a few words they identify what a person in

**TABLE 5.1**

Effects of a Broader Span of Control on the Layers of Organizational Structure

| Organizational Layers | Span of Control | | |
|---|---|---|---|
| | 5 | 7 | 8 |
| 1 | 5 | 7 | 8 |
| 2 | 25 | 49 | 64 |
| 3 | 125 | 343 | 512 |
| 4 | 625 | 2,401 | 4,096 |
| 5 | 3,125 | 18,807 | 32,768 |
| 6 | 15,625 | 117,649 | |
| 7 | 79,125 | | |

a managerial position is responsible for. A few descriptive words suffice. Long texts are unwelcome. They provide excuses for escaping accountability and they open a way allowing each manager to step on the other managers' toes.

The *span of support* rests on two legs. One is structural (Section 5.1). The other is functional; information technology being an example. The functional requires determination of numbers, kinds, and locations of all types of facilities directly connected to management support made available to meet goals and requirements associated to management responsibilities.

A useful approach in estimating support requirements is to make a precise evaluation of the efficiency with which existing facilities are being used, establishing the level of desirable upgrades and goals for various management functions. Particular attention must be paid to the weaknesses in present practices, such as

- segmented approaches, without taking into consideration company-wide needs, and
- lack of advanced technological solutions able to meet anticipated or changing managerial and professional information requirements

Decisions concerning the span of management are so important that they constitute a vital element of the corporate planning process. Plenty of companies have encountered organizational problems associated with obsolete structure. One of the major mistakes made in regard to the span of management is that companies are inclined to allow to change not as a result of a rational study but in response to ad hoc immediate problems.

Quite to the contrary, as the previous sections have explained, structural plans should be designed for dynamic growth and be able to provide both efficiency and flexibility in regard to operations. The tendency of organizations to become rigid and fail to be adaptable to constant changing circumstance is one of the most acute problems of business and industry.

In conclusion, the task of organization planning with span of management as focal point, is essential to determine managerial, professional, administrative, and service positions as well as interrelationships among them necessary to meet operational requirements.

It is well to remember that structural planning is not something abstract or foreign to the firm's everyday experience; rather, it is a commonplace requirement of practically everything we do. The best way of looking at decisions regarding the span of management is as those putting order into an otherwise chaotic situation. Solutions must be lean, because a lean organization is one of the great principles of modern management, which can serve as guide in structural planning.

## 5.4 STRUCTURAL PREREQUISITES FOR GLOBAL BUSINESS

Globalization sees to it that a company has to determine which of its products—actual and potential—can stand against competition from low wages countries. And the competitive advantage will be innovation or quality, since competing on cost will be a helpless task.

Once this basic decision is made, the company has to select and train management and professionals able to do the job at international scale under an established industrial strategy. This is true even for companies who have had international facilities, because the nature and thrust of their activities have changed considerably with globalization.

To better appreciate the sense of this reference, we should briefly return to the period between the two world wars, when exchange limitations and import restrictions made it desirable for some companies to establish factories abroad, but this happened on a rather limited scale and local manufacturing plants did not necessarily bring with them a change in marketing methods.

In most cases, the market to be served by a manufacturing plant was the geographic area defined by the borders of the country in which the plant was located. This locally operating company set up a selling organization, usually of small scale patterned after the sales organization of the parent firm with the objective to provide sales coverage only to the local market.

After World War II, however, the fact that in several cases because of lower wages local production was most cost-effective than that of the parent company provoked considerable interest at headquarters in elaborating an integrated production-and-sales network, which in turn posed organizational challenges such as

- rethinking the company's supply chain,
- coordinating domestic production planning with that of foreign factories,
- organizing and applying uniform quality standards for all factories, and

- establishing a coherent and comprehensive any-to-any program, framed through structural solutions

Such structural prerequisites must respond to management's role in global operations. This often starts with an evaluation of the investment in capital and skill. If management chooses to invest, then the extent of this investment must be determined; a decision involving a number of factors, among them

- size of the market in which the investment is made and of other open markets that can benefit from it
- nature and extent of existing local competition, as well as that from other global companies, and
- potential customer acceptance of the company's current product(s) as well as need for new ones to permit a sound market positioning

Among themselves these three bullets underline the need for a thorough market and product study as well as for defined global structural planning to develop a solid organizational framework. In addition, companies operating internationally find that to survive within a fiercely competitive market, their whole management philosophy must change. Other factors concerning decisions about the globalization of company operations are as follows:

- Political and economic atmosphere within the foreign country
- Regulations, determining whether there must be any local ownership
- Considerations of tariff, import quotas, exchange restrictions on the remittance of profits and, most evidently, taxes

Taxes can have a major impact on decisions to build, and invest and operate. An example is provided by size of newspapers.

Broadsheets, characterizing the larger size newspapers became popular after the Tories introduced a stamp tax in Britain in 1711. This taxed newspapers per sheet of newsprint. As an aftereffect, papers were published as single sheets with huge pages. It took almost 300 years for the broadsheet size, which became traditional, to change and this is proof of two things:

- The power of taxes to influence behavior
- Delays associated with changes of behavior once a certain de facto standard takes hold

In terms of international operations, a serious pitfall into which many firms fall is to take for granted that conditions influencing business abroad are bound to have rather minor variations from standards prevailing at home. Another pitfall is to let the foreign business be handled through "a foreign trade department" renamed as "international division" with little integration, if any, within the mainstream of management thinking.

The better managed multinational companies treat their foreign business as a major issue in itself and not as a minor division of a domestic unit—a process requiring significant structural choices. IBM did so in post-World War II years when it established the IBM World Trade Corporation.

Furthermore, efficiency in management decisions requires that the executives of the foreign enterprise get more autonomy than do managers of the domestic units. This is necessary for practical reasons since there is not, nor can there be, the same degree of executive management supervision over day-to-day business abroad that is practical with respect to operations in the homeland. In other terms

- the men in foreign markets must have the power to act speedily,
- but still corporate management has a most effective instrument through which to exercise supervision over all operations, foreign as well as domestic—the profit targets and budget.

It is the responsibility of the global firm's CEO and CFO to examine, amend, and approve the annual budget of each foreign operation. This should evaluate the projected expenditures for the year ahead and also judge actual performance by its yardstick—sales. This budgetary approval is the foreign executives' marching orders. The *plan vs. actual* evaluation is the measurement of how well the job has been done.

It is no mere coincidence that global industries and financial institutions have been moving toward the sensible system of appraising managerial performance by comparing actual results against verifiable quantitative and qualitative goals.

Profit and loss statements permit the evaluation of executive performance but executive performance itself requires standards that should be universal in the global enterprise.

- A fundamental consideration is the recognition that the most direct and efficient of all controls is the control of management quality.

- One of the most important challenges facing the multinational firm is an adequate development and evaluation of its managerial personnel.

Human resources policies (Chapter 6) and issues relating to organization and structure correlate. As a chief executive of a leading manufacturer commented, "Basically, we operate under the principle of national managers reporting to the president and executive vice-president of the company.... The man running the U.S. group has full responsibility for the operation of the group, and so on. This in many ways answers the question concerning the hiring of local nationals, which we favor highly. Their promotion possibilities are excellent in as much as each national manager is on a par with the other."*

As far as human resources are concerned, due consideration should be given to activities such as matching people with job requirements independent of country of origin. Centralized leadership of human resources does not contradict the spirit of effective decentralization.

Last but not least, to remove elements of doubt the global enterprise should be keen to audit to appraise the quality of its managers and professionals. How well does a person plan, organize, staff, direct, and control activities under his authority? Top management must put itself in the position of elaborating and asking fairly objective questions that can cast useful light on these issues, while avoiding confusion arising from overlapping functions and responsibilities and costly duplication of effort.

## 5.5 THE JOB OF DOWNSIZING

*Downsizing* is a relatively new concept in the business world and it means the paring of people, products, facilities, divisions, and/or areas of operations. Cutting down in size was once thought a drastic step to be taken only when hard times hit or, when all other cost-saving efforts were exhausted. Employees regarded the first waves of downsizing as they would any other tragedy: an accident waiting to happen to someone else.

But after the wave of mergers and acquisitions of the 1960s when the most disperse businesses came under the same hat—examples being ITT and Litton Industries—downsizing became a sort of a policy. Even companies not

* From a personal meeting.

under dire financial or competitive pressures started to look at it as a move that would jolt managers from their complacency and leave the company

- leaner
- meaner, and
- more fit to compete in an increasingly tough marketplace

Slowly but unmistakably, managers, employees, and the companies they were working for have come to realize that downsizing is here to stay. It is a corporate prerogative used not once but repeatedly, in good times and bad. In fact, this is what business and industry, with a world-wide excess capacity of nearly 50 percent, today needs on a global scale; and the automotive industry is a case in point.

Another reason for downsizing is that corporations are facing relentless investor pressure to rethink their profit line and business portfolios. Management seems ready to comply because the company's profitability also reflects in its bonuses. Spin-offs are a way of doing so, particularly if a subsidiary no longer fits in long-term growth plans. That vague concept of a division which used part of the aggregate but no more does so, includes

- a subsidiary that is not meeting expectations, and
- a unit that does not fit within the latest strategic plan.

Slimming down is also a way of capitalizing on tax advantages. Curiously enough, in several countries the laws allow to preserve more value in the typical spin-off than outright sale of the same business. Shareholders tend to benefit because some 80 percent of the shares of the subsidiary are distributed to shareholders. When structured this way, the spin-off is tax-free for both the investors and the company.

An example from the go-go 1980s is W.R. Grace who divested to equity holders of National Medical Care its kidney dialysis division. Prior to the spin-off, W.R. Grace received a $3.5 billion bid for the unit from director L. Hampers. But an outright sale would have resulted in an estimated tax hit of $850 million to $1 billion. A recent example is Xstrata of Zug, Switzerland, one of the world's biggest mining firms, born out of the initial public offering (IPO) Glencore's coal mines in 2002. It has grown spectacularly since, and in 2010 it was worth $52 billion.

The latest shifting of assets between the former parent and the spin-off came on March 5, 2010, when Glencore exercised an option to repurchase

Prodeco, a Colombian coal mine. Glencore had sold that mine to Xstrata in 2009 for $2 billion to raise the funds to participate in Xstrata's rights issue.

The skill of Xstrata's top brass has been in buying lower-quality mines and running them more efficiently. The company achieved this greater efficiency with an unusually slim and decentralized management: Xstrata employs only 50 people at its main offices in Zug and London. (The latest in store in the Glencore/Xstrata love affair is that they may be merging.)

Not all spin-offs work out, however, to the stockholders' advantage. Sometimes downsizing is a way of correcting past mistakes. In December 2001, faced with a torrent of red ink from its (unwise) expansion into the broader range of financial services and acquisitions that came along with it, Zurich Financial Services (the former Zurich Insurance) floated its third party reinsurance under the name Converium.

Analysts did not like the terms of this flotation because they were complex. Converium took over most of Zurich's reinsurance business through a quota share retrocession agreement, excluding business written under the Zurich Re brand name before 1987.

- Converium was given responsibility to administer the transferred business.
- But Zurich Financial Services remained liable to the original cedants.

According to the terms of the spin-off, by managing third party retrocession related to the transferred business Converium was responsible for the credit risk for uncollectible reinsurance. The new company could offset reinsurance balances it owed to Zurich against the funds withheld account, and it could request cash advances from Zurich if it had eligible losses from a single event or series of linked events.*

Among other special terms to the Converium flotation, Zurich capped the new company's losses from the events of September 11, 2001, at $289 million. It also agreed to bear any relevant uncollectible reinsurances. Zurich covered as well all Converium's net losses above $59 million related to business ceded by workers' compensation insurer Amerisafe. As these references show, spin-offs are

- a legally and financially complex business, and
- some are also loaded with potential liabilities

* Not related to the quota share retrocession and exceeding $25 million up to a maximum of $90 million.

This is evidently to the advantage of the former parent whose liabilities are downsized. Some people call the process of downsizing "thinking about shrinking," but this in no way changes the fact that it means cutting down—often with a sharp knife. In a number of cases, this is not a choice. It is a necessity, both in the business world and in public administration.

On March 23, 2010, the citizens of Detroit packed into a large auditorium in the city's mid-town. Charles Pugh, the leader of the city council, was present, and Dave Bing, the mayor, was due to give his first state-of-the-city speech. Lecturers and listeners were aware that harsh realities had produced new radical thinking. In fact,

- dramatic steps were discussed seriously, including plans to close dozens of schools, cut services, and transform the landscape, and
- in counterpart, the urgent tasks were to create jobs, cut crime, and clean up a fiscal mess that threatened to bring the city to bankruptcy

The plans for downsizing the City of Detroit involved major changes in its landscape. The mayor said the plan was to demolish 3000 homes in 2010 and 7000 more by the end of his term. And this would be only the first step toward restructuring Detroit.*

Government sponsored or financed entities are also faced with downsizing problems. In the United States, the two white elephants Fannie Mae and Freddie Mac have as yet escaped downsizing,† but the British Broadcasting Corporation (BBC), which since the end of World War II became indistinguishable from the British government's landscape, did something unprecedented. On March 2, 2010 it volunteered to cut itself down to size.

The broadcaster decided to abolish two digital radio stations, shrink its website and spend less on imported shows and sport. Mark Thompson, BBC's general manager, said the networks he is running will abandon the goal of turning out more and more programs to suit every taste. The decision is an important change from the policy of all-out expansion that guided the BBC, but it is pragmatic particularly when judged against the fact that

- in 1981, BBC1 and BBC2 mustered among themselves over 50 percent of the British share of TV viewing,
- while by the first decade of this century this share shrank to well less than 30 percent

* *The Economist*, March 27, 2010.

† Still surviving at taxpayers expense.

At BBC, as in plenty of other companies, many jobs have been redundant. To downsize their personnel requirements without hurting their ability to deliver, companies invest in technology. Knowledge engineering constructs, for example, perform functions formerly the remit of middle management layers and "assistant to" personnel.

No matter what the reason for downsizing is, there seems little doubt that careful, well-planned pruning can transform a company into a better competitor. The job is tough, but it offers rewards. Rare is the company that does not suffer from at least one unneeded layer of management, and rare is the nation that does not suffer from overstaffing of public services across the board.

Speaking in straight business terms, successful downsizing requires selectivity and an overall strategic plan that looks at the work and structure of the organization. Control systems, such as careful budgeting and hiring practices, have to be put in place to keep a company at the right size. In addition, the preservation of gains from downsizing calls for a system that pays attention to the longer term.

## 5.6 REENGINEERING THE ENTERPRISE

The deep economic and financial crisis of 2007–2010, as well as the number of big banks that went to hell, particularly in the United States and Britain, have posed the very intriguing question of whether very large companies with tremendous overhead and burdensome bureaucracies are able to manage themselves. Contrary to the large hierarchical structure of the big corporation, an entrepreneur can

- make quick decisions,
- rapidly respond to changing market pressures, and
- ignore Wall Street's need for ever-increasing quarterly profits

Because smaller fast-moving business units fit better the perspective and demands of a dynamic market, there is a growing interest among private individuals in buying up bits and pieces of large concerns. In essence, this means that the buyer must have seen something in a company up for sale that some classy management had not.

Sperry's Remington Shavers has become a classic case. Its asset was that it has been a leader in electric shaver technology for over 30 years. It had a solid international network for sales and maintenance; featured manufacturing facilities world-wide; and featured few competitors (Philips, Braun)—hence it had a steady market. People still need to shave and there has been no product substitute to the electric shaver.

Sperry Rand decided to sell even though the electric shaver division better than broke even. Bad management had much to do with this disinvestment. At a first time, Sperry tried to expand and make the electric shaver division a growth industry. This failed: electric shavers are a stable industry—not a growth industry. There have been, as well, technological problems of the false kind: a too fast change of models created technical issues where they did not exist.

In 1979 Remington Shavers was sold to Victor Kiam for $25 million, but the reasons were unclear. Kiam, a marketing executive with experience from Lever Brothers, identified both the unwisdom of continuing "innovation" and its limits:

- The policy of fast (six months to one year) technological change was making obsolete the bestsellers—and development costs were high for no returns.
- Because of emphasis on technology, little thought was given to price differentiation. "Past models" were just dropped from the product line.

Kiam changed that policy: New models, such as the "3000," were priced at $40. Elder models were reintroduced and, given the depreciation of equipment, were priced at $20. In 1980, within a year of the new policy, pretax profits amounted to $2.6 million—more than 10 percent of the purchase price.

American Safety Razor (ASR) is another case. The company made Personna and other lines of blades. Philip Morris had acquired ASR in 1960 and continually tried to turn the small blade company, with annual sales in the $40 to $50 million level, into a formidable competitor of Gillette, the giant that dominated the (at the time) $800-million market. But

- Gillette was a too powerful rival.
- Philip Morris never came up with a cohesive strategy for ASR that worked.

As a result, ASR became only marginally profitable, and its sluggish performance had psychological consequences. Philip Morris began trying to sell ASR in 1974, but only to a buyer that would promise to keep on the employees. Different well-known firms, including

- American Home Products
- Coca-Cola Bottling
- Sterling Drugs
- British Match

as well as others looked and walked away. Bic Pen, already challenging Gillette in blades (as well as lighters), agreed to buy ASR in 1976 for $19 million. The deal, however, fell through after the Federal Trade Commission objected on antitrust grounds.

Then Philip Morris began liquidating ASR. It cancelled most advertising and dismissed about 100 salesmen, 80 percent of the force. Within two years, ASR would have been totally out of the shaving-blade business had not John Baker (the ASR president) and eight of his fellow executives bought the firm for $16 million, putting up $600,000 of their own funds and borrowing the rest.

The then president Philip Morris told Baker that he was making a mistake. "The day of the little entrepreneur is over," he remarked. But John Baker decided ASR could prosper by concentrating on some of the smaller segments of the blade business Philip Morris had ignored because going after a small piece of the blade market did not have any appeal to the big parent.

In the few years after the management buyout, the Personna line, which accounted for 28 percent of sales in 1979, contributed less than 10 percent. By contrast, blades for industrial uses, like scraping windows, and private-label blades exceeded 20 percent of ASR's sales volume—four times as much as in Philip Morris days.

Subsequently, in the early 1980s, ASR purchased a private-label soap business from Procter & Gamble, another indication of the change in strategy in terms of product line. As ASR became a comfortably profitable business, razor-blade sales hardly increased at all, but pretax profits were way up—more than 400 percent the level under Philip Morris.

Helena Rubinstein is still another example. In 1973 the company was sold to Colgate-Palmolive for $142 million. For the next four years its operations were profitable, but by 1978 Helena Rubinstein was losing money and Colgate-Palmolive decided to sell. A Japanese company Kao Soap negotiated to buy the perfume firm for $75 million, but the deal did not go through. The French L'Oreal offered $50 million. This deal did not go through either.

Finally, Burack-Weiss got a deal difficult to refuse. It only laid $1.5 million in cash, with the balance to be paid over several years. After the purchase, Burack-Weiss carefully studied the product line of Helena Rubinstein and its operations.

- The good news was that 25 of the firm's divisions were profitable.
- The bad news was that five countries accounted for 90 percent of the losses, with the deeper trouble in the United States, England, and Japan (representing 38 percent of sales).

But the damage was not irreparable, as a relaunch of certain Rubinstein products proved. Consumer reaction was very favorable. Burack-Weiss made a thorough market research, which established that mass-market products were inconsistent with the Helena Rubinstein image. The new parent

- sold the mass product brand names, collecting royalties, and
- focused its next moves by concentrating on high quality

Another rigorous study by Burack-Weiss established that the U.S. and English factories were inefficient and were operating below capacity. Hence, management closed them down, brought production to a continental European factory and exported to the United States and England. In a short time both the British and American operations were in the black. By 1981, Helena Rubinstein was solidly profitable.

What Burack-Weiss did in reengineering the Helena Rubinstein operations is a first class example on what can be achieved in enterprise through internal restructuring. The company boosted its operations by selling underperforming businesses and those that did not fit into its strategic plan. It also reengineered the product lines that it retained and paid great attention to both factory and market profitability.

# 6

# *Staffing and Directing*

## 6.1 HUMAN RESOURCES STRATEGY

The organizational assets we today call *human resources* have been classically known as "personnel" or *staffing*. The goals of the staffing function are those of recruiting qualified people, watching after their job performance, steadily improving competence, countering human obsolescence by providing lifelong training and assuring career planning.

It is no mere coincidence that at top of this list of a company's important staffing functions and in the management of human resources is recruiting qualified personnel. The reason is self-evident:

- *If* able-brained people don't enter at the base of the organization,
- *then* there will be no qualified persons to lead the enterprise when they reach the top, and the company will fade into oblivion.

Assigning qualified people at each position, developing a valuable human inventory, and helping in improving their knowledge and experience is a "must," not an option. Lifelong learning makes a big difference in the quality of staffing (as the preceding chapters already brought to the reader's attention). Buddha said that we should live as if it is the last day of our life, and we should learn as if we live forever.

Let me add some more references connected to recruiting. It is no coincidence that it is not enough that the CEO sends his director of personnel to interview graduates of the best universities, in order to find the right entrants into his company's base. Before he does so, he must ask himself: What do I want my company to be in the next 15 to 20 years?

- Where will it be competing?
- Which human traits will be needed for such competition?

- Which are the services on which it should be strong?
- How is it going to produce and deliver these services?
- What should its future executives be able to reach?

The common thread of these queries is "What kind of personalities and skills will bring us from here to there?" The answer defines the company's staffing strategy. Neither past policies nor models and stereotypes can provide answers. Able answers will come from the men at the top—not from books or machines.

In 1917, George F. Baker, president, CEO and chairman of New York's First National Bank, invited Jackson Eli Reynolds to join his inner circle. This proposal came out of the blue, and Reynolds, a corporate lawyer and Columbia University professor, responded with a spontaneous burst of laughter.

"Have I said something funny?" the 77-year-old Baker asked. "It sounds to me," Reynolds responded, "as I am expecting a telephone call any moment from Paderewski asking me to play a piano duet with him." "Banking is not as hard as piano playing," Baker assured.* He might have added that, by contrast, the selection of the next CEO requires the skills of a maestro.

Getting ready to lead a business takes a great deal of preparation. Meyer Guggenheim started out like thousands of other mid-nineteenth-century immigrants to the United States—which means he built his business from zero. Guggenheim was not only a hard worker but also a driving, inquisitive man who swiftly turned nothing into something.

In 1880, using stock market gains and earnings from a well-established embroidery business, he turned two flooded mines in Colorado into silver and lead money makers. Then came the challenge of ARASCO, the copper mining and smelting entity operating in the western United States, Canada, Chile, and Bolivia.

Meyer Guggenheim planned very carefully his family's switch from embroidery to copper. The way Bernard Baruch relays it, he sent all his seven sons to learn the copper business.† Training is where he made his first important investment, while biding his time for the right moment to take over ARASCO.

---

* James Grant, *Money of the Mind*, Farrar Strauss Giroux, New York, 1992.

† Bernard M. Baruch, *My Own Story*, Buccaneer Books, Cutchogue, NY, 1957.

Success, says an old proverb, is what happens when preparation meets opportunity. As far as family expertise in copper mining and smelting was concerned, the Guggenheims were prepared when the opportunity came. In 1900, ARASCO's cash-strapped investors begged him to take a majority interest and save the firm. In the decades that followed

- the seven sons built on the business, expanding into tin and nitrates for fertilizers, and
- when America went the big way for electrification, the price of copper soared, and the Guggenheims were major beneficiaries

This is the *golden rule* in staffing, as far as preparing the CEO's succession is concerned. First, the strategic choice, closely followed by training, training, and training—getting ready to catch the opportunity as soon as it comes. It is no wonder that staffing is such a crucial element in the corporate planning process.

An integral part of building up human resources is the determination of the qualifications, numbers and kinds of people to meet the organization's current and future requirements. As the preceding examples underlined, first comes the definition of corporate goals then the human resources qualifications challenge shows up—and it continues looming as one of the more critical problems in facing the future, for it is quite clear that

- There is and will be a marked shortage of adequate, qualified talent at practically all organizational levels.
- Systematic planning of human resources is essential if a company is to secure and maintain product and market leadership.

Such systematic planning involves several basic steps. For example, an analysis of the various positions in the organizational structure, to determine the types and characteristics of people needed by current and projected business. This should be done position-by-position, as incumbents move on or retire.

Since no organization is better than its staff, these determinations are criticized in defining the firm's future course. Needless to add that emphasis should be placed on special skills and knowledge required by each person and each position (see the discussion on Responsibilities, Indicators,

Objectives in Chapter 5). The better the staff, the more likely the company is to

- accomplish its goals,
- develop an effective operating organization, and
- achieve the programs, projects, and objectives it has set for itself

The persons with brilliant ideas whom they employ enable them to develop innovative products and concepts far more effectively than those of their competitors. The imagination of Walt Disney created a company that for several decades has been without parallel or rival. GE was built on the extraordinary ingenuity of Thomas Edison, the Ford Motor company on the concepts of its eponymous founder. The spirit behind IBM was Thomas Watson, Sr., and Akio Morita of Sony occupies a similar place in the annals of modern business.

The contribution of first class human capital can best be appreciated if we account for the fact that in a globalized economy the success of established firms is generally based on their depth of managerial *and* technical expertise—also on their marketing skills and distribution capability, which must be steadily upkept. Short of that, they will be overtaken by competitors.

## 6.2 THE LAWS OF HUMAN RESOURCES ARE ASYMMETRICAL

Today, capital buys brainpower, not just equipment. When in June 1995 IBM put down $3.5 billion for Lotus Development, it was not just purchasing the company's software products but, rather, the skill and vision of Lotus Notes architect Ray Ozzie and scores of other creative minds employed by the firm.*

Because innovation is based on out-of-the-box ideas that are the driving force in high tech, many in the technology industry's workforce (whose own compensation is often linked to the success of their companies) understand that their future is nonlinear. It is tied to the willingness of

* Eventually Ray Ozzie left, becoming Microsoft's top thinker and senior executive.

investors to

- take a chance on a good idea, and
- earn an impressive rate of return for that risk

The management of settled, traditional companies not only needs to recognize this sea change, but it should also develop a policy—or, more precisely a culture—that reflects it. No company can afford to remain transfixed by the vision in the rearview mirror, because this leads straight to the precipice.

As long as they remain oblivious to the mind-boggling changes taking place during the last 30 years, senior executives can position neither themselves nor their companies to face the challenges of this century. Knowledge is productive only if it is applied to make a difference. This is true all the way

- from knowledge planning, and therefore education
- to the results we wish to obtain in a way that is measurable, transparent, and visible

Moreover, imagination and knowledge can make real contributions only if they are clearly focused. This requires both the right education and hard work. It also calls for systematic exploitation of opportunities as they present themselves, balancing

- the short-term outputs in the sense of deliverables, and
- the longer-term, knowledge-based evolution, which distinguishes an advanced society

Most critical to this process is the conceptual change, which permits to mobilize the multiple pieces of knowledge one possesses, interconnecting these pieces into a flexible, restructurable system. As far as the management of human resources is concerned, the ability to communicate, inspire, and guide professionals to create a more highly motivating work environment is cornerstone to success.

For a person, a company, and a nation the deeper roots of education are not only the investment with highest return, but also what in a modern society characterizes what in the best possible way we call human resources. Theoretically, this is what is supposed to differentiate the Western economy from that of developing countries. But is it practically so?

The answer is not unequivocal because existing evidence demonstrates that as the educational system expands to cover more and more of the young generation, the quality of its deliverables becomes lower. It looks *as if* the knowledge the educational system can produce is something resembling a fixed quantity—or, alternatively

- discipline breakdown with mass education, and
- going to school ceases being synonymous with learning

"The laws of education are the first we receive…they prepare us to be citizen," said Montesquieu, the great French writer and philosopher, several centuries ago. "If the nation in general has a principle, then the parties composing it…will also have one."* Most unfortunately, mass education is no more cultivating knowledge, and this has significantly reduced the level of know-how by spreading it too thin.

One would like to think that this is the wrong hypothesis, but existing evidence says otherwise. Moreover, the effort many young and elder individuals put into education is far from being deep and rigorous—with the result that the laws governing human resources are uneven. This is magnified by the fact that the educational system is under stress, and the choices young people make in their field of study are questionable.

- Soft subjects like sociology, literature, and political science are easier to study than hard subjects.
- But what the labor market of industrial nations needs is engineering, physics, chemistry, biology, medicine, mathematics—the hard sciences that promote technology.

This discrepancy between science and technology orientation demanded by a knowledge economy and the availability of graduates in soft subjects is another reason why the laws of human resources are uneven. In the aftermath, a university degree no more guarantees employment.

The way an article in *The Economist* had it, the story of British higher education is less about expansion than inflation of hopes which are not sustained by qualifications. University degrees mean less and less and there are more and more of them. The rot, *The Economist* says, set in 1992, when the

* Montesquieu, *De l'Esprit des lois*, Classiques Garnier, Paris, 1961.

Conservative government allowed a poorly controlled upgrade of locally based institutional teaching, relabeling them "universities." That act

- created a panoply of new academic courses, many of dubious merits, and
- did away with a vital pillar of the higher education system by striking out the difference between the purely vocational further education colleges and the fully academic universities*

The aftereffect is the rise of unemployment among graduates. This is very negative because not only is contribution to and satisfaction from one's education being significantly diminished, but also finding employment is becoming much more difficult than it should have been. Take Spain as another example.

Prior to the real estate boom of 2002–2007 and after the bubble burst, Spain was and has been a country mired in recession. Unemployment hovered around 20 percent with an upward tendency, and stiff labor market laws carried significant costs for the country's economy. This has been well known, but, whether center-left or center-right, successive Spanish governments have refused to take unpopular measures.

Instead, successive Spanish governments have developed the habit of showering public money on any problem, largely money they got from the EU in a bargain that had (incorrectly) classified Spain as an underdeveloped country. This proved to be a big negative for the country's human resources. Today, Spain is suffering from an expensive, poorly educated labor force—and coming up from under means reforming both the

- dysfunctional education system, and
- disastrously managed labor market

Collective-bargaining has been another scourge. Its inflexible rules mean that wages for the country's labor aristocracy are rising even as prices are falling. Since Spain, a Euroland member, can no longer devalue, its businesses are becoming even less competitive—while in 2009 its budget deficit headed for 12 percent of GDP and in 2010 the country has been hovering at the brink of bankruptcy.

* *The Economist*, January 24, 2004.

In conclusion, education is a "must," but the development of human resources has prerequisites. To start with, the educational system should work like a clock. This is not happening today as the mass effect and the trend toward soft subjects has disastrously weakened the educational process.

Then the people themselves should want to be truly educated and develop worthy careers. If they drop out of school, or even if they go to school as a pastime, there is no wonder that the laws of human resources will be uneven—they will be unemployed and live miserable lives. "Chance favors the prepared mind," said Louis Pasteur, the great French chemist.

## 6.3 MANAGING THE HUMAN RESOURCES

Companies favor a centralized function of personnel management, both for reasons of better top management control and for more effective coordination. Though some basic responsibilities of human resources may be left to the subsidiaries, the central director sets the rules and, typically, he is instructed to pay a great deal of attention to the operating units'

- staffing policies,
- quality of intake,
- upkeep of know-how,
- synergy of skills, and
- career management

With global corporations the multinational central human resources staff must clear up with the subsidiaries personnel objectives, each important case being judged in its own merits and not accepted because it happened to be in "this" way in the past. The corporate executive vice president (EVP) of human resources is typically responsible for screening the local director's ability to handle the personnel he leads, and has available people of managerial timber. To do so in an able manner, he must know how to deal with

- intrinsic motivation, and
- extrinsic reward

He or she must also be knowledgeable on how to create incentive systems that keep professionals on the go by offering them interesting work, good wages, and a career development commensurate to each one's contributions.

In collaboration with local management, the central director of human resources must have a system able to keep promotions that bypass technical experts from being interpreted as signals a given person is not good enough to be a manager. In its heyday, IBM had gone around that problem by creating a distinct class of "IBM Fellows." Jim Bachus, who had developed Fortran, was its first member.

Just as important is having a method for better integrating newly hired engineers and scientists into the organization. Another responsibility of the corporate EVP of human resources is to resolve personnel differences at headquarters and within subsidiary organizations, helping people to understand their roles in important projects requiring high-performance contributions.

Top of the line organizations provide plenty of evidence on the importance of avoiding being boxed into protective positions. Corporate personnel management must show the subsidiaries how to overcome the unique problems associated with the performance and productivity of individual professionals. Also how to enhance their creative abilities.

All this adds up to a demanding task because the approaches best adopt to motivation change over time, and so do the methods needed to maximize productivity (Section 6.7). Part of the reasons for this change are the sources of tension that exist between organizational and professional demands. While trying to leverage the effective contribution of their technical and managerial people, they don't always understand about

- the relationship between motivation, innovation, change, and uncertainty,
- the difference between uncertainty and risk being assumed,
- ways existing to reduce uncertainty, that allow to introduce change effectively, and
- how to surface problems early on, which is one of the best ways to avoid, or at least relieve, subsequent stress

Speaking from personal experience, the foreign subsidiaries of global firms tend to value stability, not change. Even at headquarters, protecting old ideas and techniques becomes a main preoccupation of middle management strata (Chapter 5), whereas creating new opportunities is way down their line of priorities.

One of my professors at UCLA had taught his students that in life problems are opportunities, and what I have just mentioned is an opportunity—not a problem. It is an opportunity whose solutions will also bring other benefits because preparing for change means upgrading and renewing the skills of employees—as a way for improving the firm's scientific and marketing knowledge base.

One of the most important challenges for the human resources boss is overcoming complacency and beating inertia, which is a prerequisite for acting on new opportunities. Measurement is at the heart of any improvement process. *If* progress toward better human resources cannot be measured, *then* this is evidence that they are not being improved. But to be successful,

- the metrics must be set a priori, and
- the measurement must begin at the outset of the improvement program, not in the middle of it or near the end

Even better, such measurements must be accomplished by the participants to the improvement program—that is, by the working group. It is the managers and the professionals themselves who should see to it that this process does not become bureaucratic, there is evidence of an *esprit de corps*, and no turf fights or squabbles break out.

This essentially amounts to saying that human resources need leadership. It is the personnel division's responsibility to instill the leadership culture. Objective measurements provide better insight and early warning signals lead to significant competitive advantages. Worrying about protecting today's position is less important than creating tomorrow's leading edge.

As the foregoing references document, the management of human resources is not just one of personnel selection and disciplinary action, though discipline itself is very important. Vital responsibilities are leading, providing for steady education and career planning as well as of assuring a steady supply of experienced executives who are needed to run the organization.

The corporate VP of human resources is also responsible for promoting among the multinational staff specific responsibilities assigned to men and women of different nationalities, and to assure they are given equal opportunity. In this mission he should be careful to assure that promotions are based on performance. Among other vital responsibilities in this job description are

- establishing standards for and approving proposals by subsidiaries on personnel policies, and
- assuring that the company has the human resources it needs to respond to (or even initiate) drives for market leadership

Where this adds up to is making manpower management a most critical part of overall international business planning, execution, and control. The job is demanding but it can be effective if the major factors involved in the management of manpower are determined and translated into detailed requirements. Companies with experience in this domain suggest that the task of the international human resources staff is one of high-level planning and guidance, assistance to the subsidiaries, and coordination among

- foreign subsidiaries, and
- various groups and departments of domestic operations

A great lot of foresight and practice is required for appraising managerial performance at headquarters and subsidiaries level, as well as for setting performance standards for effective manpower planning. The latter also involves a steady management manpower inventory, covering all positions in the global organization chart and outlining the criteria by which will be judged success and failure in terms of human resources.

In conclusion, planning assumptions must be established to identify expected organizational changes through improved processes and methods, including the acquisition of skills. At international staff level, the personnel EVP is expected to compare existing and anticipated requirements and define new management positions in terms of

- responsibilities,
- functions, and
- duties to be performed according to plan

Without any doubt assuming responsibilities for human resources management and development for all subsidiaries is a demanding job, which includes the realization that board members and the CEO must also be involved in fundamental decisions. Modern requirement for skill and know-how is more than what a classical personnel department can offer.

## 6.4 THE ACT OF DIRECTING

Directing means many things. Among them tactical decisions made day-to-day within a strategic plan; delegating, while retaining responsibility; assigning jobs to be done and seeing that they are done; administering, as well as coordinating and motivating the way discussed in Section 6.3.

Many people think that directing is a rather bureaucratic activity. Though in cases it might be so, this is not by any means a general truth. Directing does involve, however, administrative chores, painstaking attention to detail as well as overcoming personal and other differences. The best performers in this mission encourage independent thought, stimulate creativity, and carefully control deliverables.

A steady part of the job of directing, hence the manager's job is to relate the efforts of subordinates into winning combinations, as well as to get the most out of them (see also Section 6.7 on productivity). More than anything else, in performing his duties the manager must be in charge of the chain of command:

- Assigning authority commensurate with responsibility
- Assuring the attainment of goals by carefully supervising projects under his authority
- Demanding accountability from each one of his subordinates

The deliverables the director expects from his subordinates must be high quality, within budget, and respecting the time plan. He is expected to perform not only this part of the functions (according to his job description), but it is also an assignment whose results will characterize how well the director discharges his or her responsibilities.

Clarity and detail in the order the director gives is at a premium. It is a golden rule of directing that the orders the boss gives should neither be too general nor abstract. Orders must be clear and unambiguous. Anecdotal evidence says that during World War II in the Eastern Front headquarters, in East Prussia, Hitler was using as interpreters of the general staff's orders some of the most stupid soldiers the Wehrmacht could find for him. The rationale was that

- *if* a stupid soldier in the quietness of headquarters could not understand the order,
- *then* a general in the front, under steady stress from fighting, would not understand it either

I had a friend, CEO of a major company, who after he issued an order at an inflection point or under stress conditions, would call his EVPs in a meeting and they would have to explain to him what they understood the order said. Notice that these were not flat orders but rather firm guidelines from the boss. As Alfred Sloan once remarked, flat orders destabilize the organization.

Let me repeat this reference. The best executive orders are simple in character, to a large degree quantitative, and easy to understand. Apart from other advantages, this assures that the extent to which they are being applied is measurable and controllable. In addition, the orders given by a director should be

- directly related to daily activities of his subordinates, and
- integrate within the goals set by the board, as well as more general orders given by the CEO

Goals that do not integrate, or even worse, contradict each other and/or the company's strategic guidelines, create inefficiencies in execution. In turn, organizational inefficiencies tend to increase frustration. They deteriorate morale amongst the most ambitious members of management and they destabilize specialist groups.

Therefore, manpower planning, recruitment, and selection (Sections 6.1 to 6.3) are inseparable from the function of directing a given line of activity. The boss should also get involved in programs targeting his subordinates' training and career development; closely cooperating with the human resources division in matters concerning *his people* relative to personnel decisions are as follows:

- Job description*
- Career planning

* Including responsibilities, indicators, and objectives (Chapter 5).

- Salary structure
- Work standards to help judge performance
- Tools and policies for auditing performance

It is the manager's responsibility to allocate missions to human resources under his watch, and define timeframes necessary for their accomplishment. Indeed, all assigned tasks should be time-phased and individual persons should be made responsible for the accomplishment of each task within explicitly stated deadlines. Steps must as well be taken to provide for regular reviews of progress being made—something comparable to what we do through *design reviews*, in connection to engineering projects.

It is unfortunate that schools don't teach since early age that time is a resource like money, and in many cases time is more precious than money. Timetables and efficiency in doing once a given job correlate.

In his book *The Merchants of Debt* George Anders gives an excellent account of KKR, the investment firm, and its culture noting "... the KKR partners could count on a third big source of investment profits: increased efficiency. Almost every big company was hidebound by excess overhead and bureaucracy, Kravis and Roberts came to believe."

A particular sore point was that in many companies' headquarters staffs were too large and ineffectual. "I call them the people who report to people," Kravis often remarked. If most big companies could whittle down their overhead costs, the KKR partners believed, profitability could be greatly improved without the need for any sales growth. "Companies build up layers and layers of fat," Kravis told a newspaper editors' luncheon in 1989.*

Henry Kravis has been perfectly right. Corporate entropy leads to a steady increase of administrative costs; and costs matter. As Chapter 5 demonstrated, lean organizations are better positioned to survive and they are characterized by few management layers. Fat is the enemy of health—whether we talk of individuals or of organizations.

Sometimes because incompetence, nepotism, and poor management reach a level that is no more affordable, management is obliged to act, but in its effort to reduce cost it cuts muscle rather than fat. On March 3, 2010, TV5, the French television channel, had a report from Quebec which stated that in this century the number of people working in public works

---

* George Anders, *Merchants of Debt*, Basic Books, New York, 1992.

increased by 700 percent. By contrast, the number of engineers grew by only 20 percent, which meant that

- projects are understaffed in engineering skills,
- but they are overstaffed in stage hands and in bureaucrats—with disastrous results on quality and efficiency of the work being done

In the early 1960s, Professor Parkinson published a small but juicy book, *Parkinson's Law,* which perfectly demonstrated that expenses expand to exceed budgets. To prove that a bureaucracy tends to grow exponentially Parkinson used statistics of men in active duty and in offices, in the British Admiralty before World War I. During World War I the fleet in the Admiralty and the office workforce expanded. But after the war ended, the fleet shrank while the number of the Admiralty's bureaucrats continued rising.

There are also some other principles of executive life connected to the function of directing, to be brought to the reader's attention. With the exception of the CEO who, in several cases, tends to be a *prima donna*, directing is a relatively low profile job. Managers, their professionals, and employees should nevertheless appreciate that administrative duties can be creative and a track record does not just happen without a lot of hard work.

In addition, other things being equal, the more skilled and the more trusted the assistant to whom the manager delegates work, the "luckier" the manager is going to be. The same is true of vendors and consultants. "We have never been inclined to give any undeserving stranger a free ride, and we will never change our mind about that," said Sam Walton.*

## 6.5 CONCEPTUAL AND DIRECTIVE PERSONALITY TRAITS

Typically in literature as well as in many peoples' minds, the terms *executive* and *manager* tend to be used interchangeably. This is general practice, but it is not precise. The executive is a policy maker, which is not necessarily the manager's case, though there are exceptions to this statement.

* Sam Walton, *Made in America: My Story*, Bantam, New York, 1993.

For instance, people at senior level are both executives and managers. Independent directors (members of the board who do not work for the company) have executive but not managerial positions. Typically, a manager is a person internal to the firm for which he or she works. With the exception of internal board members, who (as stated) are also managers, a classical example being the CEO, the difference between executives, therefore policy makers, and managers, occupying themselves with day-to-day activities, may be much deeper than what the labels are suggesting.

First class policy makers are *conceptual* people, whereas managers have experience in *directing* their subordinates. This is a soft dichotomy as a great deal of difference is made, as well, by personality characteristics.

Figure 6.1 presents the four quarter spaces of what are known as personality characteristics (not to be confused with intelligence, hence with IQ). Cognitive contingency self-rating models have been designed to uncover which of the quarter spaces shown in this figure is the stronger one in a given person, which is the next in strength, and which the weaker. At the root of this chart lies the fact that

- people are either left- or right-brained, and
- they are recognized as being primarily thinkers or doers, represented (correspondingly) by the upper and lower layers in Figure 6.1

Great mathematicians are in the top left quarter space, but those who develop profound theories have both analytical and conceptual capabilities (hence upper layer in Figure 6.1). Master of industry, the so-called tycoons, have both conceptual and directive characteristics (diagonal in Figure 6.1). Generally speaking

- the left brain is characterized by effectiveness, and
- the strength of the right brain is broader vision and shareability

A person cannot excel in everything. His or her personality characteristics are weighted. One of the best self-rating tests lets a person assign grades to each quarter space, to the total of 300 points. Let me repeat that the resulting decision style has nothing to do with intelligence; it is the person's own pattern. Figure 6.2 shows the results of a test. The person who took this decision style test is predominantly analytical with his next strength being directive—hence, left brain.

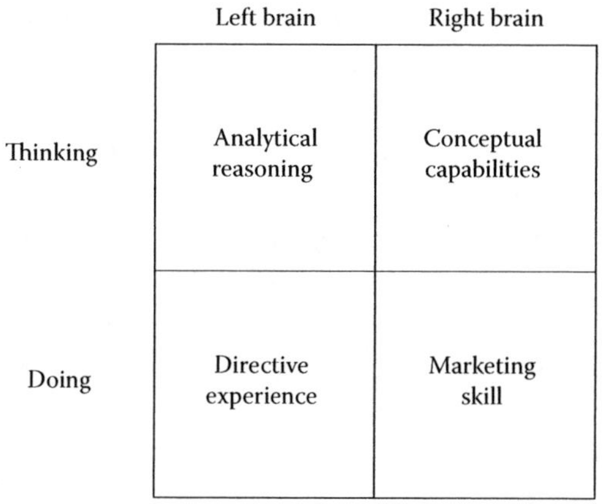

Left brain
Right brain
Thinking
Doing
Analytical reasoning
Conceptual capabilities
Directive experience
Marketing skill

**FIGURE 6.1**
The four quarter spaces in which fall the individual personality characteristics.

Least important
Back-up
Dominant
Very dominant
0
10
20
30
40
50
60
70
80
90
100
110
120
130
140
Analytical
Directive
Conceptual
Behavioral

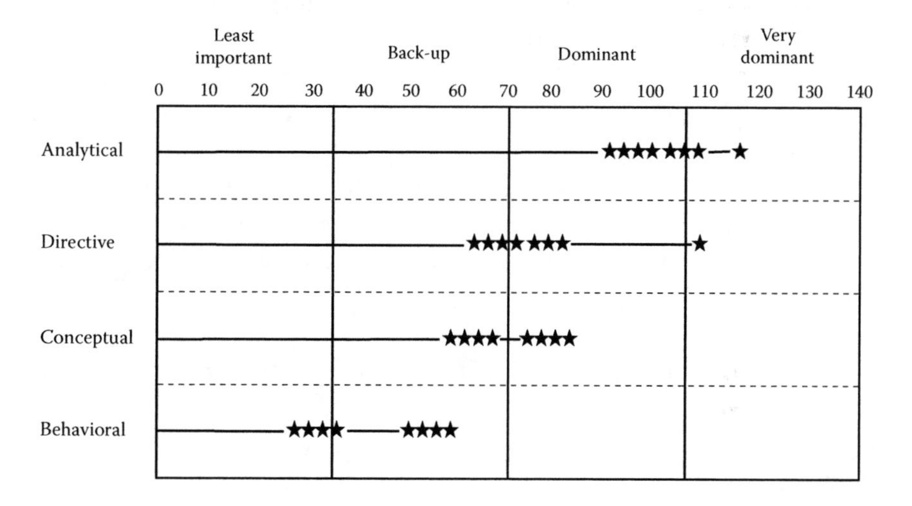

**FIGURE 6.2**
Decision style pattern of eight persons assigned to tests. The all-important conceptual skills are way behind.

Other studies on personality traits have focused an individual's ability to meditate. Their rationale is that there exist several beneficial aspects associated to meditation. These include

- better focus,
- control over emotions, and
- reduced levels of stress (Section 6.6)

There is as well a link between meditation and brain structure. In a relatively recent project at UCLA, the researchers found significantly larger cerebral measurements in meditators compared with the control group composed of nonmeditators. These include larger volumes of the right hippocampus and more gray matter.

Because the areas of the brain with increased gray matter are closely linked to emotion, the researchers conclude that these might be the neuronal underpinnings that give meditators the outstanding ability to regulate their emotions, and allow for well-adjusted responses to whatever life throws their way.*

It is proper to admit that we don't know so much about how the brain works, even if it is becoming recognized that this is of importance in the able management of human resources. "Personality" traits and behavior characteristics, like meditation, help a great deal in the proper assignment of people to jobs and managerial positions.

Another advice on sound management, in the sense of directing, has been given by Laotse, the Chinese philosopher, circa 600 BC. Laotse said "If you hire people better than you, you prove you are better than anybody." It is regrettable that very few managers really heed this golden advice. One of the reasons is the defect in business school education; one which has significantly grown over the years rather than being swamped. Most often

- business education builds egos,
- instead, it should teach humility and, with it, the strength to come up from under—therefore to succeed

What business schools should be strong at, but aren't, is the art of negotiation, building a team, finding a middle ground and solving problems. In business there is a tremendous need to influence and create relationships, to go beyond the next moment—but this is not part of current business education.

Although quantitative methods that are widely taught are important, they are only one of the aspects of managerial education. "Management as a profession of scientific, clear and antiseptic methodology is irrelevant to the needs of today's organizations," says David Noer, professor of business leadership at Eton University. "Many new managers, particularly

* *UCLA Magazine*, January 2010.

newly minted, inexperienced MBAs, come to the business world with the expectation that

- It is a place of rationality,
- Subject to objective analysis, and
- Thoughtful, quantitatively based decision making."*

Noer adds that business schools will be well advised to rectify the lopsided imbalance of head skills with an equal dose of heart skills. Managers need to connect to people, particularly when employees are in the midst of a crisis of identity and purpose. In fact, not only employees but the executives and managers themselves may be under stress and quantitative methods are impotent of pulling them out of the hole they have built for themselves.

## 6.6 MANAGERS WORKING UNDER STRESS

Stress is the body's natural reaction to events or activities that one's subconscious finds threatening or challenging. Essentially, it is a protection mechanism that enables the body to prepare for a fight or to run away from a physical threat, provided such threats are short-lived. Also provided one appreciates that under certain conditions stress is healthy (more on this later).

According to some opinions, stress is a subject *à la mode*—not an unavoidable consequence of working under pressure or of confronting difficult and complex situations. Others, however, believe that stress is being created by the dynamics of present-day business: high-risk urgent conditions, associated with an inability to manage the perception, if not the fact, that one is under constant pressure.

According to this latter school, stress becomes a major problem when the threats are emotional, intellectual, and long lasting. The problem arises when stress turns into being chronic, with the result a feeling of helplessness over day-to-day events and adversities that impact on how one works. Identifying stress requires

- taking a close and objective look at every event and activity during the work day,
- analyzing one's approach to the problem and to life at large, and

* *Financial Times*, April 12, 2010.

- evaluating in an objective rather than in a subjective manner the current and further out outlook

Indeed, in terms of further out outlook much depends on how one responds to events that come along. Inability to be in charge of a situation involving uncertainty, or to the demands of a managerial job, means that he or she internalizes responses that the body interprets as stress. This leads to *transference*. The tendency of individuals suffering from stress in one aspect of their lives, is to take it out on an innocent party or in an unrelated part of their lives.

Sometimes managerial stress is the direct result of having a very demanding boss. When he was chairman and CEO of Occidental Petroleum, Armand Hammer required that members of the board of directors, most of whom were employees, give him signed, undated resignation letters that he could use if they tried to vote against him. At about the same timeframe, senior ITT executives were regularly grilled in large meetings with CEO Harold Geneen.

Something similar has happened with large military projects when a strong personality is in the picture. To test how candidates held up under pressure, Admiral Rickover often relied on intimidation. Some officers said they were seated on a straight-back chair, one of whose legs was shorter than the others. Wobbling on the unbalanced seat, they underwent examinations that could last for several hours.*

But not everyone was under stress. Many young officers appreciated the training, and they knew Rickover chose well in drawing from the cream of the officer corp. As Fortune relayed the story, to serve under the admiral has been, "like having your ticket punched for success" said one Rickover alumnus.† Because of Rickover's training, private industry has eagerly recruited nuclear Navy men.

- Private industry held Rickover and his brand of personal management in high esteem but also in awe and fear.

To build the U.S. nuclear fleet, the admiral had to compel shipyards and contractors to adopt unprecedented high quality standards for the

---

* At about the same time, the 1960s, a similar test at IBM was to invite a manager, senior salesman, or applied science engineer to give a lecture to a group of top brass executives. The lecturer was provided with flip charts whose holes did not fit the nails of the flipchart frame, and pencils that would not write. The rationale of the stress test was to see how a person reacts to unexpected adversity in form of a group of his superiors.

† *Fortune*, November 1976.

fabrication and assembly of nuclear equipment, and for the hulls that carry the reactors.

- Rickover's power over ship contracts, and the settlement of claims for overruns, made him someone not to be crossed.

People who were involved in the build up of the U.S. nuclear submarines fleet say that stress was often associated to a Rickover inspection visit, which sometimes took place at night. These inspections sent waves of apprehension through the yard's naval and civilian personnel.

Once when Rickover boarded a submarine tender at General Dynamics' Electric Boat facility in Groton, Connecticut, to pass the night, he found his cabin telephone out of order. This brought the ship's executive officer running to the admiral, who roared "What kind of goddam repair ship is this where even the phones don't work?" Stress or no stress, it is *quality of deliverables* that made up Rickover's fame.

- His ships worked better than he had promised.
- They were built on time and on budget.

Quite unfortunately, this is not the stuff taught at most business schools—or at least it is not taught in a way that has lasting impact on students. Here is a quotation from a letter entitled "The Green Consultants' Contribution" written to *The Economist* and published by the magazine on November 11, 1995:

> (Our) company enlisted the guys from Harvard—management consultants with brains but no scar tissue. One-piece kids in three-piece suits, someone called them…
>
> One late night, our day having been extended by the necessity of attending yet another management seminar, somebody remarked a newspaper article which claimed that the cost of maintaining a convict in the California prison system was equal to that of putting a kid through Harvard Business School. Whereupon someone else remarked that the world would be a lot better off if we took all the kids out of Harvard and put them in prison. It would have been a solution of practically equal cost.

Nobody ever said something like that for Rickover's graduates. Instead, there was praise. The admiral's testimony before congressional committees was always eagerly awaited. It came replete with attacks on the Pentagon: "that kingdom of illusion." Anecdotal evidence suggests that one day a child asked the admiral why he went on—did he not see his mission was hopeless? Rickover is said to have replied "In the beginning I thought I could change men. If I still shout, it is to prevent men from changing me."

## 6.7 PRODUCTIVITY

The clock is ticking for the manufacturers and service companies whose productivity is below industry standards but costs stand above those of their competitors. Today productivity gains are more critical than ever before because globalization saw to it that prices are coming down steadily and margins are being squeezed.

The formerly secure base domestic market of many companies has been hit as well by deregulation, for not to mention strikes. Said a senior executive of a major airline, in the course of our meeting, about an ongoing strike and the strikers "Their victory would be a defeat for the company's wage-earners, cabin personnel included, as it can sign the airline's death warrant."

Survival requires greater efficiency, a mission that shows up time and again. Personal productivity is a worthy goal like physical fitness or a healthy diet—but like physical fitness it calls for change in habits.

- In the longer term producing more in the same unit of time is the key to rising living standards.
- The trouble is that in the short term there is a tension between efficiency and jobs, because people believe that this endangers their employment—and sometimes it is that way.

The contrast is presented by America and Germany in one side, France and Britain in the other. The former two have gone on a diet; they squeezed extra output from their workforce. France and Britain, by contrast, have opted to contain job losses at the cost of lower productivity. This made their industries uncompetitive in the globalized market place.

Both for companies and for national economies, it is easy to single out productivity growth as an important gauge of industrial health. Nothing affects more long-term living standards than improvements in the efficiency with which an economy combines labor and capital—a statement equally valid when applied to an economy and to a person.

Typically, productivity growth is calculated by dividing total output by the number of workers, or the number of hours worked. This is admittedly an imprecise standard, which can be manipulated. But as long as it is uniformly applied it provides a basis for comparison.

In principle, even if metrics and benchmarks are not as precise as one wants them to be, they help in identifying a drive—for instance in productivity, which expresses management's decision to win. One of the mechanical/electrical engineering firms, I was consultant to, nearly halved its direct labor costs after benchmarking against its immediate competitors. Another cut the number of intermediate shop floor inventories three-fold following a benchmarking exercise against a competitor's plant.

It is, however, true that all by themselves benchmarks don't tell the whole story. At times, knowing where to start in productivity improvements may be daunting, particularly if management is

- afraid to study its rivals and analyze its weak points, or
- is unwilling to spend money on restructuring the organization, shop floor

Numbers talk, but sometimes the message they give is of an unsatisfactory kind because they may hide facts that are qualitative. It is always perilous to try to compress complex problems into single statistics; 2009 productivity figures in the United States and EU provide an example.

In America, unit labor costs for second and third quarter of 2009 highlighted how aggressive companies have been in containing costs by forcefully maintaining a trend toward greater productivity. Unit labor costs have been down six percent on a year-on-year basis, as hours worked were cut (but with the result of a corresponding sharp increase in unemployment). Aggregate hours worked were reduced by seven percent.

Not so in the EU. From mid-2008 to mid-2009, labor productivity for the whole Euroland economy shrank by four percent. Indeed, in parts of the manufacturing industry the drop in productivity has been nearly 14 percent, though in services the corresponding number was only one and a half percent over the same timeframe.*

To say the least, at first sight these statistics look strange. To understand what's behind them one should account for the fact that in continental Europe there is no "hire and fire" as in the United States. Labor laws make

* European Central Bank, Monthly Bulletin, October 2009.

firing, and therefore also hiring, a complex business, and many companies chose to keep their workers and office employees underemployed—rather than take the road to expensive, massive layoffs. It's a balancing act that numbers alone cannot describe.

This does not change the fact that when the numbers are good, management is proud of its productivity achievements. The statistics shown in Figure 6.3 come from the 2009 annual report of Eli Lilly, the pharmaceutical company, which takes pride of this fact by stating "In 2009, we continued our focus on productivity." The metric being used is *revenue per employee,* which increased seven percent to $540,000 in 2009.

Seen from a social viewpoint, productivity growth is definitely welcome both in a developing and in a developed society. However, when saturation in employment possibilities has been reached, as is the case in most of the Western countries, imagination is required in order to face the social challenges that this saturation poses. Center-left and left political parties cry loud that production for the sake of output is no longer the crux of the day because

- increased productivity leads to increased output
- increased output requires less employees
- the trend to less employees increases unemployment, and
- increased unemployment leads to ever accumulating problems

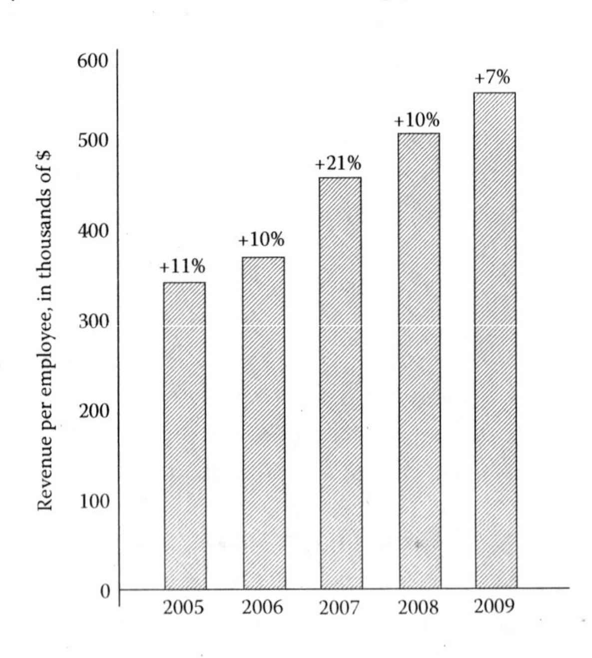


**FIGURE 6.3**
Increase of revenue per employee over five years at Eli Lilly.

At the same time, however, the strength of any national economy is its output. Other things being equal, lower productivity has a very negative effect on costs and higher production costs reduce competitiveness and therefore shrink the pool of employment opportunities.

The reader should also appreciate that, traditionally, the concept of productivity and its statistics reflect labor's results. Matters are much more complex when it comes to the productivity of executives, managers, and professionals.

For managers and professionals, the best productivity tool is the one that increases decision power by means of experimentation and the testing of hypotheses through easy to use but reliable tools. In one of the leading banks I was consultant to, the board developed a system which permitted that complex investment proposals, which took up to three man-months of a professional's work to prepare, could be produced in two hours through interactive analytical finance. Subsequently, by using the latest high-tech, this was reduced to a fraction of an hour.

Simulators increase our ability to access and manipulate the rich information we obtain from sources inside and outside the company. Information technology, when properly used, helps in freeing highly paid individuals from routine work permitting them to concentrate on knowledge production—precisely the task for which they are paid. The making of accurate models and simulators, however, is a high art—and the school system produces very few of these artists.

# 7

# *Management Control*

## 7.1 MANAGEMENT CONTROL DEFINED

To achieve its objectives, an industrial firm, merchandising enterprise, service company, or financial institution needs at top management level, an environment in which planning and control activities flourish.* A well-managed organization is distinguished by its orderly way of operating, and planning is cornerstone to this process. For its part, control action is an integral part of sound governance. A formal system of planning and control

- permits delegation with commensurate responsibility,
- coordinates and unifies action throughout the organization,
- structures the progress of accountability, and
- keeps management informed of whether or not policies are observed and plans are executed

Shortly after Alan Mulally was named Ford's chief executive (September 2006) he organized a meeting of his senior managers whom he questioned how the business under their authority was going. The answer he got was: Fine, fine, fine. "We are forecasting a $17 billion loss and no one has any problems!" the new boss exclaimed.†

As the discussion went on Mark Fields, who ran Ford's operations in the Americas, admitted that a defective part threatened to delay the launch of an important new vehicle. "Great visibility," Mulally said. In the following months a new culture took over and four years on, Ford is making good profits. Revival began with the executives' willingness to recognize and discuss their problems. That's what management control is all about.

* The process of planning and control can be structured in a variety of ways but, invariably, the goal is to assure that operating units—both line and staff—are committed to achieving target goals.

† *The Economist*, December 11, 2010.

The unifying action of management control benefits the firm by focusing attention on key areas, permitting the effective delegation of authority and facilitating the assessment of individual performance. However, while necessary control systems and procedures are not enough, management must still set aside time for evaluations and control decisions.

In terms of management control CEOs should be wise enough to follow President Lincoln's advice: "My mind is like a piece of steel, very hard to scratch anything on it, and almost impossible after you get it there to rub it out." Lincoln added "Always bear in mind that your own resolution to succeed is more important than any other thing"*—and control is cornerstone in sustaining successful resolutions.

Furthermore, control information must be both cost/effective and of high quality. Its quality is dependent on the soundness of the basic data used for purposes of evaluation. To ensure effectiveness, a control system should focus on the key components of profitability within operational plans and develop sensitivity analyses in connection to each key operational element. Implicit to this statement is the fact that management control is no different than that of any feedback:

- Drawing attention to variances from agreed-upon targets,
- Calculating the likely impact of such variances, and
- Providing information to assess the likely effect of a corrective action

Control activity will be valid only if it operates in a way fully uninhibited and transparent. Also if planners and managers properly appreciate the aforementioned basic premises as well as the need for an uninterrupted and unbiased feedback. Figure 7.1 presents in a nutshell the framework of management control.

Effective control must be exercised at both strategic and tactical levels. This is not the usual case. An example of strategic failure in mergers is the 1991 acquisition of NCR by AT&T. The formerly proud phone company was struggling for a new identity—and growth strategy—in the wake of long-distance deregulation. Its top brass believed that telecommunications and computer technologies were converging and tried to capitalize on that hypothesis:

- AT&T looked at NCR as promising acquisition candidate, as it made a profit in computers.

* Emil Ludwig, *Lincoln*, Grosset & Dunlap, New York, 1930.

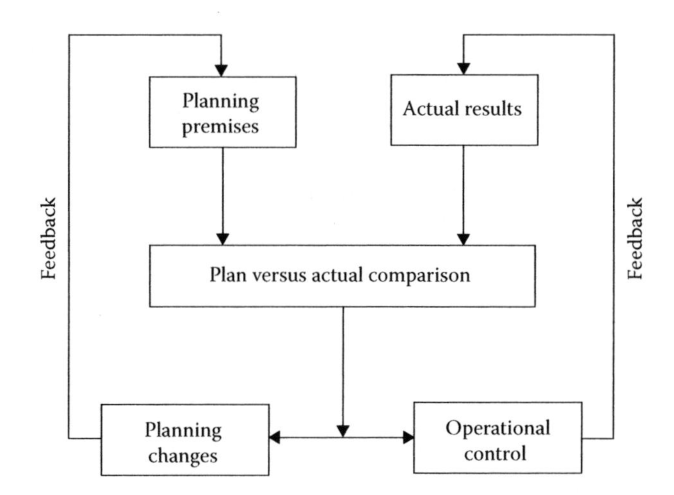


**FIGURE 7.1**
Control action is activated by organizational feedback after plan-versus-actual comparison.

- The telephone operator's board thought NCR would work in synergy with AT&T, since both companies shared not only some common vision but also a white-shirt style.

Contrary to these superficialities, however, a due diligence process, which was done too late to have the effect of avoiding a bad merger, brought to light serious problems. Engineers from AT&T's Bell Labs assessed NCR's technology only after the acquisition, while they should have done so ex ante. When they did the technical audit, they discovered substantial differences between

- AT&T's switching abilities and basic PC technology, which significantly reduced the synergies expected from the acquisition, and
- as AT&T did not care to find out, in personal computing NCR was little more than a me-too assembler; the company essentially was a mainframer

As subsequent studies demonstrated, even the hypothetical cultural similarities between AT&T and NCR proved superficial. The same was true of management philosophy. NCR operated in a highly centralized way. AT&T was decentralized and its subsequent attempts to flatter NCR's hierarchy backfired. Moreover, AT&T's management control was in shambles. Only 5 years down the line, after suffering heavy losses, in 1996 AT&T spun off NCR.

As this example from two well-known companies brings to the reader's attention, although the process of management control is indispensable it is not popular among executives. Many are reluctant to listen to the bad news and criticisms *internal control* often brings, let alone to act upon them (Section 7.3). "I don't want to hear that" is a sort of keyword greasing the wheel of poor governance. Adverse information is eliminated as a factual report advances up the management chain. When this happens

- bad news never reaches top management,
- senior executives don't have a chance to hear adverse opinions, and
- the company goes from greatness to rough times, or altogether out of business

Whether in private enterprise or in public administration, performance by the top executive and his immediate assistants makes the difference between success and failure in the short, medium, and longer term. But when control information is suppressed, the CEO's eyes and ears remain shut.

Because of this, in organizational life famous blunders happen at all times, in all fields and all issues; something similar is true of lapses of memory. In the transport industry, for instance, most of the fatal air crashes involving major airlines occurred on takeoff, and each involved lapses in pilot attentiveness shortly before he pushed the throttles to the wall.*

Whether we estimate the potential of an up or down movement by the market, evaluate projects, work with pesticides, drive cars or pilot aircraft with hundreds of people in it, we don't always pay attention to what we are doing. Therefore, we should not be surprised of very negative consequences.

Such a failure is increased multifold by the lack of feedback. Therefore, management control is a fundamental organizational function. It is not something done just to kill time or make people unhappy.

Management control is predominantly a people-based process. Since organizations are made of people, we must understand the way in which people behave, act, and react.

Francis Bacon wisely suggested four centuries ago, in 1597, that if one works with another person he must either

- know his nature or fashions, and with this lead him
- or his ends, and so persuade him
- or his weaknesses and disadvantages, and so awe him
- or those that have interest in him, and so govern him

---

* According to the results of an investigation; from a meeting on airline safety.

## 7.2 BUSINESS REPUTATION

Rigorous internal control should span the organization (Section 7.3) and aim at providing evidence on whether management is in charge. It must also indicate whether the board and CEO care about the institution's reputation. A company's or any other organization's reputation is

- established by its behavior over a long period of time,
- closely tied to its activities, and the way it executes them, and
- can be easily upset by adverse opinion of counterparties, therefore of the market

Reputation and image are not the same issues, but neither are they independent of one another. Both need feedback and management control to be sustained. Reputation takes a lifetime to establish but can be lost within days or hours. Reputation damage is measurable as a loss of market confidence, of customers, of key personnel. A lost reputation is rapidly translated into loss of trust of staff, investors, business partners, supervisory bodies, legislatures. By contrast, a good reputation creates trust.

Because they both reflect sound governance, management control and business reputation correlate. The roots of the latter are the appreciation of the behavior of the board, CEO, senior management, all company employees, and the way the entity cares for its business partners. This care is shown in quality of its products and services, innovation, cost control, risk management, customer assistance, contribution to society, and thoughtfulness about its financial results.

Within this context, the goal underpinning increased emphasis on management control is to limit the losses from operational and other failures, by keeping the information channels and feedback arteries open. Transparency and immediate response to adversity through corrective action

- help in safeguarding the capital
- assist in early recognition of deviations from targets, and
- contribute to reliable financial reporting which, in turn, enhances the company's reputation

To make internal control more effective, it is necessary to identify and revamp weak practices, while rewarding those that are correct. The chairman, the board, the CEO, and senior management are responsible and accountable for the existence of a first class internal control system and its proper functioning. Because the arteries of the organization can be clogged, internal control must be regularly audited to ascertain its rank and condition.

When in my management audits I was finding financial books held in full compliance to rules and regulations, this suggested to me that internal control acted as the eyes and ears of the CEO and his immediate assistants. And because management control is a wholesome process that evolves over time, I have also checked whether it had evolved over time. Typically in all companies, new and old, different control practices coexist. The older have been used to address abuse, fraud, and errors. But during the past decade new goals have been added:

- Reputational risk
- Management risk
- Wider counterparty exposure
- Leverage risk
- Derivatives exposure
- A long roster of operational risks that include the old targets as well as technology risk

Regular audits of internal controls along the lines described by these bullets contribute to the company's business reputation. However, they are valid only as far as the people working for the organization continue to observe sound principles. For instance, principles connected to financial responsibility and accountability. This requires

- ethical behavior,
- independent oversight,
- accurate measurements, and
- transparency of mishappenings

Accountability for internal control should start at the board of directors. The board and the CEO must assure that those responsible for it perform their duties in a learned, accurate, and thorough way. Confronted with information brought to their attention through internal control, board members should ask well focused questions and expect to receive factual answers. Examples from a banking environment, for instance, will be as follows:

- Are our credits diversified or concentrated in a few names? In a few industries?
- How are our credits distributed by counterparty? By currency? By maturity?
- What's the pattern of our credits by credit officer? By branch? By foreign subsidiary?
- Is there a significant number of weak credits? What's the corrective action being taken?

Other internal control queries that have to do with the bank's reputation focus on its policy with counterparties. Why is *this* party dealing in hundreds of millions of dollars in swaps? Is the counterparty a steady user of over-the-counter (OTC) or is it balancing OTC exposure with that of exchange traded financial products? What's the net and gross exposure *our* bank had (and has) with this counterparty?

Questions to be asked by board members should evidently include conflicts of interest associated to the human element: Is the same credit officer always dealing with the same counterparty? Are the traders frequently repeating the same or similar patterns in trades with the same counterparty? The mind's eye in internal control is the discovery of trends and correlations that lead to

- the firm's exposure,
- dubious ways of doing business, and
- fallout on both its reputation and its survivability

An extraordinary amount of attention is necessary because the underworld, too, has adopted organizational principles. Lucky Luciano, the former New York Mafia boss, is the first on record to have replaced traditional Sicilian strong-arm methods with a corporate structure including a board of directors. He also initiated the systematic infiltration of legitimate enterprises, which over the years became a legend of sorts.

Luciano's management style was far different from that of his Chicago counterpart Al Capone, who spent more time killing than managing and doing business transactions. Indeed, the FBI has described Luciano's ascendancy as a new epoch in the history of organized crime.

- He modernized the Mafia, shaping it into a national crime syndicate focused on the bottom line.
- He put its operations under the control of two dozen family bosses who controlled bootlegging, narcotics, prostitution, the waterfront, the unions, food marts, bakeries, and garment trade.

This corporate type structure pioneered by Luciano saw a rapid expansion to the Mafia's influence, infiltrating and corrupting legitimate business, politics, and law enforcement. Its promoter lived in a suite at the Waldorf Astoria wearing elegant suits, silk shirts, handmade shoes, and cashmere topcoats. And when Luciano's "good life" ended in 1935, others took over the chief executive officer's role.

In the year 1935, Thomas E. Dewey was appointed New York City special prosecutor to crack down on the rackets. He targeted Luciano, calling him "the czar of organized crime in this city," and charged him with multiple counts of compulsory prostitution. Luciano denied being a pimp, but he did not escape the prison, though he made it out of it for services rendered in tracking down German spies in New York and in resurrecting the Mafia against the German army in the invasion of Sicily.

A reporter who interviewed Lucky Luciano in Italy in the twilight of his life asked if he would do it all again. "I'd do it legal," he replied. "I learned too late that you need just as good a brain to make a crooked million as an honest million. *These days you apply for a license to steal from the public.** If I had my time again, I'd make sure I got that license first."† Luciano did not care for his reputation but legitimate CEOs should care a great deal—whether or not they have got a "license" to act at the edge of legality and illegality.

## 7.3 THE SPAN OF INTERNAL CONTROL

The law, said Oliver Wendell Holmes, a U.S. Supreme Court Justice, is not about right or wrong or rational judgment, but a system of sanctions. The

* Emphasis added.

† *Time*, December 7, 1998.

law and its aftermath is best seen from the viewpoint of the person, who asks not what the law books say or what reasons judges give in their rulings, but only what a specific court is likely to do to him if he is caught.

This policy dates back to antiquity, having been established by the lawmakers of Sparta, in ancient Greece. Today, it fits well the concept underpinning the span of internal control—both within an organization and in society at large. Politics, however, sees to it that this law of deep personal responsibility is not properly applied. Take the financial industry as an example.

The scams of 2003–2007 that led to the deep economic crisis of 2007–2010, and the financial bubble that accompanied them, went through an array of steps: from very high leverage to the subprimes, illiquidity, insolvency, the central banks printing press and, in 2010, sovereign risk. Even with this massive (and illogical) support by the state, banks and other financial companies are confronted with a triple convergence of exposures and of uncertainties that range across

- institutions,
- markets, and
- national jurisdictions

The fact that the risk of malfeasance, including management malfeasance (Section 7.5), is so pronounced enlarges the span to be covered by internal control. To be effective, management control must be polyvalent both in regard to the functions over which it exercises its watch and in reference to the laws and regulations. In a meeting, the Commission Bancaire of the Banque de France advised that financial institutions under its authority must have the following solutions in place:

- Control their operations and internal procedures
- Measure the risks that they assume
- Document all of their transactions
- Establish, survey, and manage their exposure
- Use a well-functioning accounting and IT system

The Commission Bancaire wants to see that the solutions each institution adopts are dimensioned in a way corresponding to its size and the complexity of its business. Credit risk must be approached on both an individual and a consolidated basis. Market risk must be carefully monitored.

Internal auditing must be regularly done, and the auditing function should report to a board-level auditing committee.

The job is polyvalent. Management's attention to risks includes not just one but several factors: creditworthiness of counterparties, interest rates, exchange rates, payments and settlements, operational risks, capital adequacy, liquidity, legal risks, and the risk of maximum exposure. In its formal, official structure,* internal control must provide management with information answering not just one but several critical queries. For example:

- Are market risks maintained at prudent level and within limits?
- Are the risk metrics employed appropriate to the nature of the transactions being made?
- Are all of the company's managers and professionals fully understanding the management of market risk?
- Is the risk exposure information they get timely, pertinent, personalized, and accurate?

Other major financial events to be tracked by the internal control system are connected to the assumption of credit risk. In the classical banking business of intermediation, credit control practices are aimed at gathering information about the counterparty, as well as its financial status and behavior, thereby avoiding major loan losses. Screening for creditworthiness includes

- repayment history of each borrower
- analysis of loan losses according to management-established criteria
- examination of likelihood of poor loan mix and overconcentration of credit risk (by name, industry, topology)
- estimation of profits and losses from loans by loan type, customer class, and individual customer, and
- unearthing of unwanted practices, such as the bending of internal control to different pressures, and other failures

Each one of these bulleted items can have an impact on counterparty risk by influencing the criteria for extension of credit. Limiting exposure

* See in Section 7.4 the discussion about the *grapevine*, an unofficial information channel.

is the business of risk management, but checking on whether established rules and limits are observed is that of internal control and of auditing. The same is true of

- fraud cases connected to financial intermediation, and
- light-handed decisions on risk criteria by line management

Well-managed companies review and evaluate their systems of internal control at least annually, using people who are independent of the function being reviewed. Among the issues on which external auditing of the internal control system should concentrate is the assessment of assumptions being made, parameter values employed in the analysis of exposures, accuracy of reported information, and the methodology being used. Any recommendation for improvement should incorporate *what should be* versus *what is*.

Prerequisites to the evaluation of internal control's accuracy is the ability to reconstitute transactions and other operations, which have taken place anywhere, at any time, in full respect of established general accounting principles. For instance, the French Commission Bancaire is asking the top management of financial institutions under its jurisdiction to make sure that there is the necessary detail, depth, quality, and reliability of

- accounting and risk control information, and
- methods for evaluation and analysis of data streams from trading, loans, investments, and other operations

This requires a great amount of care by accountants and auditors. The latter have to certify the accuracy of accounting schemata, including the expertise necessary to correct deviations *if* and *when* they arise. The going rule is that auditing of accounting schemata must be done at least monthly. In addition, controls on information technology should assure that

- the level of security is acceptable from a rigorous management viewpoint, and
- the appropriate backup of information services, so that no backlogs in evaluation procedures are created by subsystem failures or overload

Increasingly, regulatory authorities emphasize the skills to be possessed by the companies under their watches. This includes both qualification of risk controllers and expertise of their employees, particularly those working on the analysis of risks. One of the prerogatives for good management is that skills and tools must be commensurate to the problems posed by

- the size of the company,
- its domain of activities, and
- the products, instruments, and services it promotes

Organization-wise, the broader the span of internal control, the more essential it becomes to establish responsibility for it at the level of board of directors and CEO. The phrase *high-level control* has been coined to describe the board's role in connection to the responsibilities outlined in preceding paragraphs. High-level control includes all controls instituted and exercised on the initiative of the board of directors, in its capacity as the body elected by the general meeting of shareholders to watch over all of the company's operations.

## 7.4 PROMOTING DISSENTION

Internal control information does not just consist of facts and figures. The latter must be enriched with commentaries on evidence, and this may follow informal channels. The company's *grapevine*, a name given to informal channels of communication, can lead to very interesting discoveries of malpractices. The same is true of *whistle blowing*. These two terms pertain to activities that are not quite the same.

- The grapevine can work both through the line of command and by bypassing it, but generally keeps unofficial information or rumors *intra mural*.
- By contrast, whistle blowing may lead to denunciations to authorities, as it has happened in the case of Madoff; in many instances whistle blowing is well documented.

Not all whistle blowing is necessarily reliable, and there are as well plenty of instances where the authorities don't pay due attention to it. As it will be remembered, in the case of Madoff, the Securities and Exchange Commission (SEC) did not follow up the information on malpractices with an investigation, as it should have done. This lack of interest is regrettable, because the information it received could have stopped the scam early on.

Although they are both outside what management books bill as acceptable practices, the grapevine and whistle blowing may bring attention to fraud or other cases auditing has been unable to uncover. A better, and more formal, method for doing so is that senior management encourages dissention—all the way from board meetings to lower levels of supervision. This way, decisions being taken benefit from contrarian opinion(s).

The rationale behind discussions where subordinates express contrasting opinions is that this helps in exploring all possibilities. Many of the masters of industry have been keen in promoting dissention, because it helps them to base their judgment on more than one voice. Alfred Sloan has been one of the best examples of this policy (more on this later). Yet, many companies make the critical mistakes of not accepting dissent. This leads them to

- blind alley in decisions, and
- dark domains or tunnel visions from where there is no exit

Dissention is a *forte* of well-managed businesses because it opens perspectives, which would otherwise be missed. An example is provided by Ray Ozzie, of Microsoft. In October 2005, he wrote a lengthy internal memo called "The Internet Services Disruption" expressing opinions contrarian to the policies of Bill Gates. In it, Ozzie analyzed how Microsoft

- was losing ground to rivals, and
- wasted opportunities to come to grips with the new market environment

Ozzie's thesis was that the company needed to change direction, to avoid disruption in its software development and market strategies. Rather than smarting at the implied criticism, Gates endorsed Ozzie's memoir with one of his own. Then, both men stood on stage in San Francisco to launch "Windows Live" and "Office Live," two free web services intended to beat Google and its kind at their own game.

A more profound example is provided by Alfred P. Sloan, the legendary chairman and CEO of General Motors. He has been one of the few top executives who not only invited but also promoted dissention as a way to challenge the obvious and test one's decisions before confirming them. In his book *My Years with General Motors*,* Sloan recounts how as chairman he always advised other board members and his immediate assistants

- never to accept an important proposal without having dissention, and
- to promote critical discussion about merits and demerits of the issue on hand

For his part, Dr. Robert McNamara, formerly U.S. Defense Secretary and president of the World Bank, advised never to go ahead with a major project unless one has examined all the *alternatives*. "In a multimillion dollar project one should never be satisfied with vanilla ice-cream only, but choose among many flavors," McNamara used to say. Had the big banks that lost a fortune with the subprimes listened to this advice, they might have saved themselves and the taxpayers from the sea of troubles into which they landed.

The failure to listen on internal dissenting voices is by more than 90 percent due to big egos. Information that might be valuable precisely because it is out of the mainstream, can be an important ingredient of successful decisions. Highly interesting alternatives and different modes of expertise are set aside because decisions have been taken without

- listening to dissenting voices, or
- examining all available courses of action, including their risk and return

"Dick Fuld ran Lehman Brothers as if he were at war. He drove the bank hard and ignored the signs of collapse,"† Andrew Gowers said in an article. "The market has a phrase for this sort of event: *the death spiral*. Creditors and trading partners take fright at a falling share price and threaten to cut off credit lines. Alarm is magnified by modern, instant communications.

---

* Alfred Sloan, *My Years with General Motors*, Pan Books, London, 1969.

† By March 2010, nearly one and a half years after Lehman's collapse, it was revealed that through creative accounting and derivatives the bank had hidden some $100 billion of debt which weighted on its balance sheet.

Fear feeds on itself and prophecies of doom become self-fulfilling. Our freewheeling, globally integrated financial markets turn out to be built on sand."*

Standing in Fuld's way by showing aversion to risk would have been fatal to anyone's career, Gowers suggests. The firm seemed not only all powerful but also very confident that the mountains of risk that it compounded did not really matter. They were simply an attractive place to make more easy money. But it was a dangerous place, too, because aversion to dissention saw to it that Lehman Brothers

- remained overleveraged, and
- became unmanageable after the market turned on its head

Speaking from experience I can assert that it is possible to find a first-class person to fill the shoes of the unusual executive who has the guts to openly state his opinion, but often he or she is sent to exile because his current boss states "I cannot listen to that person." Or, the person in question is considered to be indispensable in another job. There are only a few explanations for the "indispensable man" says Peter Drucker:

- His strength is misused to bolster a weak superior who cannot stand on his own feet, or
- His strength is misdirected to delay tackling a serious problem, or even conceal its existence

As an example on how to manage subordinates in an effective manner Drucker takes George Marshall's policy during World War II. George Marshall insisted that a general officer be immediately relieved if found less than outstanding. To keep him in command, he reasoned, was incompatible with the responsibility of the U.S. Army.

As Chief of Staff, General Marshall refused to listen to the argument: "But we have no replacement." "All that matters," he would point out, "is that you know that this man is not equal to the task. Where his replacement comes from is the next question." Marshall also pressed the point that to relieve a man from command was less a judgment on that particular person than on the commander who had appointed him.

---

* *The Sunday Times*, December 14, 2008.

"The only thing we know is that this spot was the wrong one for that man," George Marshall argued. "This does not mean that he is not the ideal man for some other job. Appointing him was my mistake, now it's up to me to find what he can do."* Out of this policy, Peter Drucker suggests, came the future generals of World War II who were still junior officers with few hopes for promotion when Marshall became the boss of the U.S. Army.

This is an excellent example on both human resources management and on how dissent can be fruitful and eventually can carry the day. At the same time, however, no matter how great may be the methodology, sound management will be a pipe dream unless the person in charge makes things happen.

## 7.5 FIRING A BAD EXECUTIVE AND SWAMPING MALFEASANCE

F. Lee Bailey, the criminal lawyer, puts in the following words his concept about the need of doing one's homework: "My experience has taught me the importance of—in fact, the absolute necessity of—thorough pre-trial preparation... Cases are seldom, if ever, won in court. They are won by the side that comes into court fully prepared because it has slaved to find the facts of the case before the trial begins."†

A similar statement is valid in regard to negotiations, investment, trades, operations, and financial matters. A great deal depends on the originality of a person's thinking and of his accomplishments—a proxy of things to come. Every initiative we are taking involves risks, and these have to be a priori analyzed as well as steadily followed to assure they don't get out of hand.

First and foremost, this is the job of the chief executive; not all CEOs, however, see it as being part of their responsibility. Many like to enjoy the perks, salary, and bonuses that go with their job—but are late to put up the time necessary for an honest day's homework. Invariably, the latter turn out to be very poor company presidents.

Management control helps in providing the evidence to rid the organization of a bad CEO. Business experience teaches that it is always tough

* Peter Drucker, *The Effective Executive*, Heinemann, London, 1967.

† D. Lee Bailey, *For the Defense*, New American Library/Signet, New York, 1975.

to get a bad CEO out of the system. The executive under fire will resist, and this will inevitably lead to in-fighting. When talking about how to get rid of a bad senior executive, Dr. Neil Jacoby, my Professor of Business Strategy at UCLA in the early 1950s, advised his students to give a bad CEO enough cord so he could hang himself.

- This, however, takes time and in the process he may destroy the company.
- Therefore, over the years well-known company bosses found more expedient ways of cleaning house.

In the early twentieth century John Patterson, then CEO of National Cash Register (NCR), fired an underperforming executive by removing his desk and chair, parking it in front of the company's factory, having it soaked in kerosene and set alight in front of the underperformer. Shortly after World War II, the then CEO of Montgomery Ward was carried out of the company's headquarters while still sitting in his chair. Armand Hammer, when boss of Occidental Petroleum, kept signed, undated letters of resignation from each of his board directors in his desk.

Edward I. Koch, the former mayor of New York, had a different policy for firing underperformers: "I just walked in and said 'You have to go,' and they said 'We are not going,' and I say 'Yes you are, it's just a question of how you're going.'"* The boss did not bend when confronted by resistance. It's an advice of which the reader should take note.

Underperformance and embezzlement are twins, one of my professors taught his students—because underperformers misuse and run down the company's assets they are assigned to safeguard. This is true up to a point, but at the same time there exist cases of outright fraud whose existence can be found in internal control's origins.

Corporate crime is not just a disgrace. It can also be an unmitigated financial disaster, and it needs no explaining that one of management control's missions is to focus on ways and means to counteract corporate crime whose most common forms are as follows:

- Theft,
- Corruption
- Accounting fraud

---

* Edward Koch, *Mayor. An Autobiography*, Simon & Schuster, New York, 1984.

The rise in corporate crime in the 2002 to 2010 timeframe stems from a mixture of increased opportunities, growing incentives and the fact that regulators turn a blind eye. Opportunities increased as companies reduced the number of people employed to monitor employees tempted to break the rules. The proportion of frauds committed by middle managers has shown a particularly sharp rise, from 26 percent in 2007 to 42 percent in 2009.*

Economic crime is in upswing, as Madoff's $65 billion embezzlement documents. This has dismal consequences for everything from corporate morale to financial performance. It also suggests that within companies senior managers should play a more active role in combating the problem, whereas at national level political leadership should make itself felt through tough supervision and a consistent effort to stamp out economic crimes through

- long prison sentences, and
- public disgrace

An excellent example of eliciting answers to critical issues commensurate to an internal control setting that went wild are Cicero's six evidentiary questions (Marcus Tullius Cicero, 106–43 BC). Paraphrased to fit a modern day financial environment, the queries the great Roman lawyer, senator, and orator typically asked in an investigation were

- *Who*, apart from the person who signed, contributed or was witness to this decision?
- *How* did the persons involved, alone or by committee, come to this decision?
- *When* was this decision originally made, and under *which* conditions?
- *Where* has been the evidence that led to the commitment being made?
- *What* exactly did the decision involve in terms of risk? Was it subsequently changed or manipulated?
- *Why* was the decision made, which precise goal did it target or intend to avoid? Was there a conflict of interest?

CEOs who have teetered on the edge of self-destructiveness because of poor or unwarranted decisions, cannot escape the searching lights of

* *The Economist*, November 21, 2009.

these queries. Through them, Cicero forced people to practically undress in public, and made those who were not under investigation to pay more attention to their future decisions, acts, and figures.

Who, how, when, where, which, what, and why are the keys of a discovery process aiming to track and measure the forces that fuel speculation, corporate malfeasance, and scams. Through them, internal control can investigate the seven deadly sins that have the power to demolish any business, in any sector, at any time:

- Greed
- Amorality
- Lying
- Deception
- Fear
- Apathy
- Arrogance

Their omnipresence was brought under perspective in an article published by the *London Times* in June 2002, about a significant rise in corporate malfeasance in Britain. In that same month, another article in the *Wall Street Journal* pointed out that in the five years that had passed since 1997 the number of civil cases in the United States involving financial violations had more than doubled. Management control is not ticking anymore like a clock.

With hindsight, an interesting reference is that an important contributor to this quantum jump in malfeasance were technological advances like the internet, which have opened up new opportunities for fraud. Another contributor is an environment that does not sufficiently discourage or penalize dishonesty. People become fraudulent because of

- financial pressures,
- the absence of limits,
- perception of an opportunity for easy gains, and
- lack of appropriate punishment, which has turned into social misconception

"As the many victims have so eloquently testified, his crimes have imposed on hundreds, if not thousands, a life sentence of poverty," said Simeon Lake III, U.S. District Court Judge in October 2006, as he condemned former Enron chief executive Jeffrey Skilling to 24 years in prison and ordered him to pay $45 million to Enron shareholders.*

A month prior to this ruling, ex-WorldCom boss Bernard Ebbers was sentenced for 25 years and began serving them in a federal prison in Louisiana. But his CFO, Scott Sullivan, cut a deal with prosecutors and had to serve only a five-year term.

In another case, Sanjay Kumar, former head of software maker Computer Associates International, pleaded guilty to charges of securities fraud and obstruction of justice in a $2.2 billion accounting scam. He expressed remorse. In November 2006 he was sentenced to 12 years in prison. Several people are however of the opinion that for every party being convicted up to a dozen others get a free ride.

## 7.6 INTERNAL CONTROL ASSESSMENT

Since the late 1990s, supervisors require that banks must have their internal control system reviewed and evaluated by the certified public accountants (CPAs, chartered accountants) auditing their books. This is expected to include assessments of the assumptions, parameter values, and methodology being used for internal control review purposes—and lead to the nonqualified or qualified opinions depending on the results.

In case of a qualified opinion, which essentially means that the auditor expresses reservations in regard to his findings, recommendations for improvement should be reported to senior management and the board; as well as acted upon in a timely manner. The regulator's position is that internal control assessment is a steady process.

Even without the regulators saying so, well-managed companies are keen to review their internal control system at least biannually rather than wait for the annual report by external auditors. It needs no explaining that internal management control reviews should be made by knowledgeable people who are independent of the function being reviewed.

The more fundamental internal control assessments follow a holistic approach. Spotty checks lead nowhere and they may be counterproductive. The examiners should start at the top: board and CEO who receive an input from internal control. This helps in their decisions on control action;

---

* *Business Week*, December 18, 2006.

the way in which they have been executed and how accurately the results have been reported.

- Are policies and procedures well documented?
- In risk management matters are the assumptions regarding exposure valid?
- Is risk information processed and reported in an accurate and timely manner?
- Are the risk control and auditing staffs up to the required level of know-how? In adequate numbers?

Special attention should be paid to the likelihood that internal controls have been lowered: Have there been any significant changes to the company's system of internal control since the last auditing review? Have the most recent review's recommendations been enacted? Are internal controls adequate in their current status? Which improvements are necessary to upgrade internal control quality?

Evidence that internal control is inadequate is provided by scandals that have been revealed and hit the public eye. For instance, one of the issues brought in perspective with the 2007–2010 economic, banking, and credit crisis is that financial instruments connected to credit risk transfer (CRT) have been subject to insider trading.

Independent assessments have suggested that one, but only one, of the reasons was that new market participants, like hedge funds and private equity firms, have access to private information but lack the internal control and compliance structure that regulators have imposed on commercial banks under their watch. Other reasons have promoted the misuse of material nonpublic information, an example being the broader availability of underlying names in a pool.

In parallel to insider trading problems characterizing the CRT market have been regulatory concerns with respect to large leveraged buyouts (LBOs), where many participants are involved and where trading activity and price movements take place in advance of LBO deals. This is a signal that some market players have more material information than others (which in a way also amounts to insider trading).

Neither is the auditing of internal control connected to accounting practices without problems. During my research, many cognizant executives commented that the assessments of internal controls as required by auditing standards is a complex job, whereas internal control testing may vary significantly from one financial statement assertion to another.

Behind these references lies the fact that standards for auditing internal controls are not yet well settled. Several auditors perform primarily substantive tests for efficiency reasons, thereby limiting the range and depth of what is necessary for discovery purposes. At the same time, the complexity that is found (at least sometimes) in internal control auditing tends to lead to some issues being misunderstood.

Auditors who respect their mission and their clients do the utmost to obtain plenty of knowledge not only about the effectiveness of internal controls, but also the weaknesses therein. Moreover, it should not be forgotten that—whether the reference is accounting books or internal control—the opinion expressed by an auditor upon completion of his or her work is an expression of

- professional skills, and
- integrity expressed through independence of opinion

Without independence of professional opinion the auditing report is as useful as feathers on a fish. The audited company should never permit itself to test hope against the facts and against experience. Several considerations are to be borne in mind in connection to the practice of auditing internal control (and the books) in order to assure that there is no misunderstanding of this function:

- The examination made by the auditor is not to be regarded as nothing more than a process of routine verification.
- A discovery process is a demanding enterprise.

While observing generally accepted auditing standards, the auditor must carefully exercise his or her informed judgment as a qualified professional person. This is the sense of the qualitative appraisal to be given by the auditor in accordance with SOX Act. It is nevertheless possible that conditions of an unusual nature, and in cases of collusion or conflict of interest, may subsequently indicate error in judgment. Alternatively, the findings might have been biased by unreliable data.

Precisely because the information the auditor is provided with might be unreliable, to be able to give an independent opinion he or she must be versatile in legalistics. Lawyers use the expression of "presumption of law," or *inference*, to mean that an opinion is sometimes conclusive but more frequently it is rebuttable.

In so far as a presumption has been enacted as stipulation in the profession's code of ethics, it has the force of professional law for the auditor. Without excluding the bearing of other rules, at top of the list in matters of the auditor's independence is the free expression of opinion

- on the existence of a valid internal control function
- on false or misleading statements related to its operation
- on conflicts of interest prevailing in a client firm, which mask the facts, and
- on occupations incompatible with control activities that result in internal control bias

In auditing accounting records, financial statements, and internal control, the course that has just been outlined does not imply the attitude of a prosecutor, but that of a judicial impartiality that recognizes an obligation for fair presentation of facts. Independence of opinion means honest disinterest on expression of his or her findings and qualification of statements. This requires

- unbiased judgment, and
- the strength to put such findings in writing

The auditor owes such impartiality not only to the firm's management, employees, shareholders, and bondholders but also to the other creditors of the business, the supervisory authorities, and those who may otherwise have a right to rely upon the auditor's report. Let's always remember that audit is an indispensable element of every effective management system.

In addition, as prescribed by SOX, management audit is distinct from the function of exercising day-to-day supervision. Internal control's audit is essentially a management audit, and therefore it finds itself at higher level than the classical audit functions such as in-depth analysis of the books to uncover the now classical cases of fraud—which always exist, but in terms of impact they have been overtaken by management malfeasance.

# Part Three

# Marketing and Sales

# 8

# *Marketing*

## 8.1 MARKETING FUNCTIONS

The first and foremost goal in marketing is to position *our* firm against market forces. To do so, we must look beyond day-to-day commitments and face the broader responsibilities for markets and sales. With globalization, deregulation, and rapid innovation, these responsibilities have tremendously increased, particularly so as survival cannot be taken for granted.

To position themselves in the market's eye, firms need to *gain recognition* both by developing products that their competitors in the industry do not have (see Chapter 11) and by being masters in public relations. Other things equal, advertising, promotions but also word of mouth (in reference to quality, functionality, affordable price) distinguish a company's products and services from the rest of the lot (Chapter 9).

- Promotions campaigns are a basic element of strategic thinking in marketing.
- They are important complements to other more classical tools such as market research (Section 8.4) and the contribution of product planning.

Marketing planning is done for the short, medium, and long range, the long-range planning focusing on the "big picture." Detailed marketing plans are short to medium range. Answers to longer-term questions require plenty of insight, as they become obvious only over a period of time.

There is always the risk that some forceful and energetic approaches may become counterproductive when the marketing manager loses sight of the longer term. As for the shorter term, some failures may not be due to a lack of hard sales efforts, but rather to the results of it. Not infrequently, it is the direction of sales that fails—and this is nowhere more true than in customer handholding. It is ironic but true that

- some sales organizations specialize in irritating the customer, and
- they don't do their utmost, as they should, in serving him

Sound marketing plans not only assure seamless continuity in sales, but are also structured in a way permitting to account for radical changes in the market's direction, or in "established" standards. They are also able to cope with the impact of a broadening geography (national and multinational) brought about by intensified trade.

Fulfilling such requirements calls for structural developments that go beyond the two-dimensional organizational chart, which we know from the past (Chapter 5). The marketing effort may involve plenty of special missions and still be thoroughly focused and able to identify, attack, and solve a myriad of customer problems. It must as well be characterized by industry-wide sensitivity to and knowledge of

- strategies of the competition
- product(s) the market wants or "needs"
- drives, location, and extent of new markets
- business-oriented analysis of what competitors (and *we*) have available to fulfill the market's wants
- steady training of the sales force, enriched with information pertaining to *ours* and competitors products and moves, and
- ample resources—both human and financial—to face marketing and sales challenges, in order to grasp business opportunities

It needs no explaining that all this requires a comprehensive analysis and design of marketing moves, which evidently presupposes ample knowledge of corporate objectives and plans. This should involve the products, sales effort, and the commercial structure itself. Based on experience from computer systems, Table 8.1 shows some salient considerations.

Crises come and go in any marketing organization and, more often than not, there are good reasons behind them—for instance, declining product appeal, growing competition, rapid innovation by competitors, lost market

**TABLE 8.1**

Product and Sales Objectives and Marketing Queries

| Product and Sales Objectives | Commercial Queries |
|---|---|
| • What's our product?<br>• What's our "next product?"<br>• Is our personnel ready for them?<br>• What's the quality of our customer handholding?<br>• What's our maintenance policy?<br>• How good is our account control? Is it satisfactory? | • What's the market?<br>• What's the market's evolution?<br>• How strong is *our* functional marketing by industry sector?<br>• Do *we* have the sales force to sell our products?<br>• Is *our* basic software still top-of-the line? Will a switch to de facto standards improve *our* sales?<br>• Do *we* have enough applications routines to sustain current market requirements? To open new markets?<br>• Are we training our salesmen and *our* clients' personnel in using *our* equipment?<br>• Which are the further requirements to support the sales effort? |

leadership, stagnant growth rates, changing marketing patterns, failure to tap new markets, highly cyclical market effects, declining management skills, overtaxed executives, and overextended capital resources.

Other reasons behind marketing adversity is loss of prestige and image, as well as inconsistencies in running the customer base. Correcting a situation before it goes out of hand requires a conceptual model for determining how well tuned *our* marketing efforts are.

One of the best in the computer industry calls knowledge and understanding of the use of the products we sell; high salesmanship standards with quantitative objectives; understanding of the user's problem and research on how *our* equipment can fit into *his* system of operations; and analysis and examination of the user's procedures to see how, by adjusting them his operations can become more productive. Three basic management duties can be defined within the overall market planning cycle:

1. *Comprehensive planning*, assured by establishing standards and guidelines throughout the organization—valid at any time and in any place.

Objectives exert a unifying influence on the implementation of plans made by different people at different times, thereby streamlining marketing and sales coordination. Even if there has been full agreement on market objectives, these may not be fully understood by all people. Quantitative standards reduce such ambiguity, and reducing them to coherent patterns provides a better frame of reference than description by words.

2. *Indoctrination and motivation* by helping the sales force to quickly understand the scope, character, and direction of marketing programs.

A well-studied and clearly stated short-to-medium range marketing plan leads to a reasonable uniformity. Quantitative sales objectives (*quotas*) help each salesman to think in terms of his own contribution to the corporate effort—to his own benefit and that of the organization as a whole.

3. *Enforcement of accountability*: When sales objectives are meaningful, precise, and tangible, they provide a workable and controllable work environment and accountability becomes generally transparent.

Quantitative objectives offer benchmarks for measuring progress, thus building a bridge between known and the unknown sales outcomes; hence the "realistic" and the "idealistic." The assessment of such objectives helps avoid organizational conflicts, which often arise because different levels of supervision are not exactly sure of expected sales results and their order of priority.

On the basis of sales objectives, marketing management sets the pattern for the future much better than based on trends derived from historical data. Objectives provide the sales force with a comprehensive picture of "things to come." They also help to visualize the momentum of the organization, its ability to meet established goals, its strong points, and its weaknesses. Attaining these objectives promotes a feeling of confidence to the company's people at a frame of reference above

- conflicting market forces,
- prevailing conditions in the industry (systems solutions, new technologies, price wars, support facilities), and
- action to be taken in regard to price policies and the drive for return on investment (ROI)*

Three levels of reference characterize the described marketing effort. The first is planning, and it is conceptual. The second is analytical, and after the facts, it is supported by statistics.

---

* ROI and market share are not the same; at times, these two aims are contradictory.

The third is that of designing prerogatives. This is creative work, and constitutes a prerequisite to implementing an effective marketing approach, including the actual management of the sales network, training of the sales force, and developing improved approaches to customer handholding—always aiming to keep abreast of market developments.

## 8.2 A MARKETING ORGANIZATION'S BEST EFFORTS

A customer-based orientation is a good policy for the marketing organization as a whole, but both the specialists at headquarters and the salesmen should be trained in establishing and maintaining it. Customer orientation and handholding require specific aptitudes, they don't come as matter of course. The players and complementarity between directives and action is shown in Figure 8.1.

Whether in this or in any other area, every marketing organization is confronted by an interplay between the planning side (typically the marketing staff) and sales proper, which constitute the executive arm, (usually working under pressure). The action of the salespeople is measurable through the results the company obtains, and that of the staff by means of the quality of directives that it gives.

**FIGURE 8.1**
A marketing organization's planning and execution of a customer-oriented policy.

Marketing efforts are being conditioned by the culture of the organization, the training of the sales force, and the way in which it executes its functions. However, the definition of marketing is not cast in iron. Coca-Cola defines marketing as anything it does to create consumer demand for its brands. Marketing management focuses on continually finding new ways to

- differentiate the firm's products, and
- build value into all of its brands

Both marketing skill and marketing spending aim at enhancing consumer awareness and impact on consumer preference for the company's products. A fairly widespread target is to produce growth in volume. For instance, Coca-Cola targets per capita consumption and share of worldwide beverage sales.

The marketing efforts of many companies concentrate on heightening consumer awareness and product appeal for their products, services, or trademarks by means of integrated marketing programs. This can be done in different ways. Many firms in IT bet on strategic alliances to implement a marketing program. To establish the best partners for such alliances, they

- start with brand positioning,
- conduct product research,
- engage in joint advertising programs,
- develop business partners communications, and
- solicit feedback from their client base

A strategy based on these five pillars can find more general applicability. To maximize the impact of its advertising expenditures, Coca-Cola assigns specific brands to individual advertising agencies. This makes feasible to increase accountability and enhance each brand's positioning.

In the general case, the creation of a planning program seldom proceeds in the straightforward, logical manner cookbooks may suggest. Its forward and sometimes backward steps are human events in which personalities, organizational tradition, and sometimes chance play major roles. Any planning program has imperfections; these can be corrected by means of

- planning feedback, and
- an uninhibited needed interaction between planning and appraisal

The close interaction of planning and review is evidenced also by the fact that some planning roles are simultaneously review instruments, such as budgets (Chapter 14), and performance standards. As it has been discussed in Chapter 4, planning must provide from the start an effective framework for review: a clear statement of objectives that should be accomplished, description of the results that are anticipated, definition of resources allocated for their accomplishment, and the names of people assigned responsibility for execution.

To learn from miscalculations (that is, to plan better), it is necessary to analyze what went wrong in the past, and this is one reason why a continuous and systematic follow-up to the planning process is so important. Without such a review, most people and organizations tend to be careless about goals, objectives, and plans. A careful review has both motivational and policing functions.

An appraisal of program effectiveness can be tested in various ways, one of the favored being the financial test. Large and complex organizations usually require, in addition, an administrative test that provides evaluation through continuing management controls (Chapter 7) that produce a regular flow of feedback information.

- Recurrent appraisals give management a detailed periodic review of selected areas of organizational behavior.
- Such reviews help ascertain the prevailing degree of efficiency in the use of plans, and also suggest adjustments necessary in the future.

A feedback process usually requires more than just the input available or generated from within the organization itself. The response obtained from clients is most critical. Just as clear goals are needed for the adequate measurement and evaluation of performance, so is adequate feedback necessary to continually appraise the adequacy, desirability, and feasibility of current objectives.

Answers given to the following seven questions help in ascertaining whether those setting marketing goals, and establishing the market planning process, have done their homework:

1. What percentage of market (and what market rank) does our company plan to attain?

This query should be answered in marketing terms for the company as a whole, by product category, as well as for how long the present responses will be valid.

2. What sales volume do the marketing targets represent by product category?

Answers must be precise, in response to queries such as, "In which markets? Individual products? Through how much added investment can the targets be increased by 10 percent? 20 percent? 30 percent? At what level of profit margin does each step-up make sense?"

3. What should be the ratio between selling costs, costs of goods sold, and operating margins?

There is no rule and no clear trend in answering this query, which is one of the most important confronting the CEO and EVP marketing. Figure 8.2 presents an example with a sample of eight well-known technology companies. At one extreme, is a well-known computers vendor whose selling costs are nearly equal to costs of goods sold. At the other extreme is another computers firm, the selling costs of which are roughly one-eighth of costs of goods sold. Operating margins, also, vary widely.

4. How can *our* company increase its influence in the market?

The answer should be specific for each market the company is active. Is the projected increase commensurate with its goals? Is profit improvement part of the marketing plan? How much can be expected in profit improvement from cost reduction? From product pricing revision? From expansion of sales?

5. Is there a timetable for an increased profitability? How much is marketing implicated in it?

Ways and means of improving profitability depend on much more than only marketing decisions. The answer to this question must specify, "When? In what product categories? By how much? How do the contemplated moves compare to those of *our* competitors?"

6. Which new investments will be required at marketing and sales side?

Marketing must study the implications for the division of contemplating corporate expansion in products and/or markets. Pertinent questions

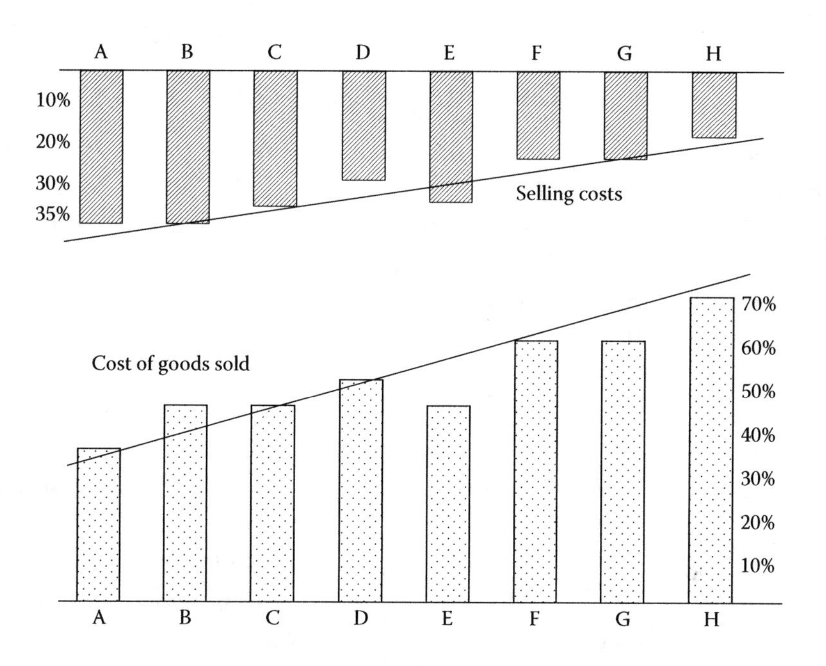


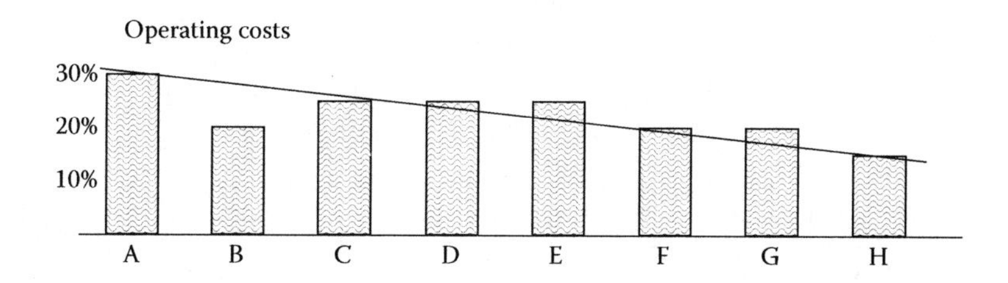


**FIGURE 8.2**
High technology companies with diverse policies in selling costs and in costs of goods sold.

are as follows: When? For what products? For which markets? What's the expected ROI? Is this above or below current results? Industry standard?

7. What kind of new skills are needed in marketing and sales?

When? How fast? At what cost? Are these costs risking to upset current profit estimates? What policies are to be followed relative to Brand name? Product quantities? Market entry costs? Permanent establishment in new markets? It's all part of a longer-term perspective in marketing and sales.

## 8.3 THE LONGER-TERM MARKETING PERSPECTIVE

Inherent to the market planning effort is the need for corporate management of client contacts, to evaluate investments in advertising, sales, promotions, and other items entering the marketing budget. Changes in strategy may dictate new strategic marketing, targets and tactical initiatives, but with the exception of inflection points such developments are fairly predictable. Still it is the responsibility of the chief marketing executive to

- search out and develop satisfactory markets for the company's products and services, with special attention to markets presenting greater potential.
- collaborate with R&D to innovate and/or broaden the product line, to facilitate further expansion both within and outside the current customers base.
- develop new distribution methods and channels where feasible, as a means of improving participation or entering into new and different markets.
- assist customers to better appreciate the company's products, prices, and value added over what is offered by competitors, and
- improve the sales figures and profits thereof, through increased efficiency of the sales force, all the way down to the individual salesman level.

These goals are longer range and most companies use not just one- but two-, three- and five-year planning periods. When the lead time necessary for the development of a product's marketing effort facility is longer than a year—as in defense, computers, and utilities—the planning period usually conforms to the period for which relatively firm contracts or commitments will prevail.

The longer-term view in marketing is an indispensable complement to the company's R&D effort. Far-out planning (Chapter 4) is 80 percent marketing and only 20 percent technology. This in no way means that what the longer view says will happen. Markets change and so do customer drives.

In the late 1960s, AT&T projected that in 10 years' time more than 50 percent of its income would come from data communications, with voice taking a secondary role. It never happened that way, wireless communications

**TABLE 8.2**

AT&T's 2009 Business Income

| Business Line | Percent (%) |
|---|---|
| Wireless | 45 |
| Wireline Voice | 25 |
| Wireline Data and Managed Services | 25 |
| Advertising | 5 |

*Source:* From AT&T's 2009 Annual Report.

being one of the reasons. As Table 8.2 shows, 40 years down the line in wireline (fixed line) communications voice and data are about equal—but by far the larger share is wireless communications.

According to the International Telecommunication Union (ITU), by 2010 there have been about 4.6 billion mobile subscriptions with penetration rates rising steeply everywhere. To my view this is ridiculous as in some countries subscriptions outnumber the population. Another important element that alters the business perspective is the merger of handys, computers, and Internet browsers (Section 8.7).

A market's explosive growth obliges that marketing planning is highly dynamic. This is, however, one of the factors affecting lead time. In the general case the particular planning period for marketing chores vary, depending upon the type of company, individual product line, and lead times associated with major capital expenditures. Although, as just mentioned, this generally ranges from one or two to five years, the planning range may also be 10 years or longer.

The longer-term marketing perspective is particularly important in industries where products compete with those manufactured by many other companies and sold in highly competitive markets throughout the world—for instance, computers and semiconductors. Another example is animal health products that confront on global basis not only the wares of animal health care companies but as well of pharmaceutical, chemical, and other firms operating animal health divisions or subsidiaries.

Other reasons, too, weigh in marketing lead times. In the electronics industry, important competitive factors include safety, compatibility, ease of use, and cost/effectiveness. Not only are the results from R&D important, but profits are also dependent on sales wizardry, as products and services compete with others already on the market as well as new products introduced by competitors.

- *If* competitors bring to the market innovative products with effectiveness or cost advantages,
- *then* our firm's offering can be subject to progressive price reductions and/or a shrinking volume of sales.

The expiration of patent protection can also have devastating effects. Eli Lilly's 2009 Annual Report makes this point, stating that "When a branded pharmaceutical loses market exclusivity, it normally faces intense price competition from generic forms of the product. In some countries outside the United States, intellectual property protection is weak or nonexistent and we must compete with generic or counterfeit versions of our products."

Eli Lilly's 2009 Annual Report further states:

> We believe our long-term competitive position depends upon our success in discovering and developing (either alone or in collaboration with others) innovative, cost-effective medicines that provide improved outcomes to individual patients and deliver value to payers, together with our ability to continuously improve the productivity of our discovery, development, manufacturing, marketing, and support operations in a highly competitive environment.

This, is, most clearly, a challenge to engineering and to marketing.

Another important duty of marketing is the protection of the company's intellectual property and trademark. This is critical to a company's ability to successfully commercialize its products and services, without having its market decreamed by unethical competitors.

Companies are busy in applying for patents relating to products, devices, and manufacturing processes—but the global market is full of counterfeits and copycats. There is no assurance that the patents a company obtains will be enforceable if challenged and global watch of counterfeits is a "must."

For instance, the marketing division of Société Suisse d'Industrie Horlogiere (SSIH), to which I was consultant, had a legal department dedicated to tracking copycats and counterfeits of Omega, Tissot, and other of the company's brands worldwide. Its activities had plenty to do with discovery of the origin of the copycats, identification of their intermediaries and customers, as well as with policing action.

In certain cases, the tracking of counterfeits can be combined with market research thereby assisting the company's efforts in identifying customer trends and drives. (Since the late 1950s—therefore, for more than half a century, market research has become an indispensable tool of the marketing effort. But not all market research efforts are effective, as we will see in Section 8.4.)

## 8.4 THE MARKETING MISSION: CASE STUDY ON WRONG-WAY MARKET RESEARCH

No two companies have precisely the same marketing structure or methods, even if they operate in the same field. Still, while the organizational chart may vary in terms of lines of authority and corresponding responsibilities, the functional aspects present significant similarities.

A great deal of the services outlined in Tables 8.3, 8.4, 8.5, and 8.6 are found in every marketing organization, albeit under different headings. The purpose of setting up these tables is to present to the reader the wider possible distribution of marketing's functionality, presented along the lines of an organizational solution adopted by a company I was consultant to the CEO. The following themes are included in each of the tables in reference.

Let me repeat that this is only an example intended to help the reader in developing his own list of marketing functions. Note, however, that some of them which used to be pillars of marketing have weakened, largely because they are used the wrong way, or assigned to nonqualified people. Market research is a case in point.

The main reason why market research fails is that its agents look at things from a twisted point of view. For instance, they ask, "Do you like this brand?" The classical market research approach assumes that the consumer is very rational, even if we know that more than 80 percent of the decisions being made originate in the unconscious part of the brain.

In addition, some of the questions asked by market researchers presuppose that the respondent is knowledgeable about the subject—which is far from being always true.

That sort of trying to extract opinions about things that have not been clearly explained to the people whose opinions are collected and distilled through market research leads to totally biased and unreliable results. It is as well a great danger not only to the concept and practice of market research—but also to democracy.

Precisely because the now classical ways and means of market research have been overly misused, and the results leave much to be wanted, there is a new species of market researchers, the *neuromarketers*. They believe their work will be especially useful for products consumers find hard to

**TABLE 8.3**

Functions of Policy Administration, Market Research, and Marketing Practices

| | |
|---|---|
| **Policy Administration** | • Contract administration<br>• Credit policies definition<br>• Exchange regulations<br>• Legal counsel<br>• Master customer files, domestic<br>• Master customer files, big foreign<br>• Master customer files, international accounts |
| **Market Research** | • Market analysis (product-oriented)<br>• Area/industry potential evaluation<br>• Business intelligence<br>• Study of competitive practices<br>• Strengths and weaknesses' evaluation of sales force<br>• Historical data analysis for sales projection |
| **Marketing Practices** | • Profit margin definition<br>• Profit margin follow-up<br>• Area development studies<br>• Potential by industry branch<br>• Rental or sales policies<br>• Product service policies<br>• Product service follow-up |

**TABLE 8.4**

Functions of Sales Management, Customer Handling, and Order Processing

| | |
|---|---|
| **Sales Management** | • Sales forecasting (by product line)<br>• Sales forecasting (by area and industry)<br>• Share of market (estimates)<br>• Share of market (objectives)<br>• Sales objectives (quotas)<br>• Sales commission decisions and coordination |
| **Customer Handling** | • Customer handling policy definition<br>• Sales performance control, by major customer and territory<br>• National accounts<br>• International accounts<br>• Direct sales<br>• Dealerships<br>• Specials and exceptions |
| **Order Processing** | • Order clearance<br>• Screening of specials and exceptions<br>• Customer notification<br>• General scheduling production<br>• Production order follow-up<br>• Cost/benefit estimates by client and order category |

**TABLE 8.5**

Functions of Sales Promotion, Marketing Coordination, and Sales Force Development

| | |
|---|---|
| **Sales Promotion** | • Advertising policies<br>• Advertising practices (public media)<br>• Direct advertising<br>• Exceptions (national and international)<br>• Showrooms and demonstration centers<br>• Sales promotion assistance—own operations<br>• Sales promotion advisory service—dealership |
| **Marketing Coordination** | • Product line "A"<br>• Product Line "B"<br>• Product Line "C"<br>• Product Line "D"<br>• Product Line "E"<br>• Product Line "F" |
| **Sales Force Development** | • Salesforce training<br>• Salesforce continuing education–sales<br>• Salesforce information meetings<br>• Sales/R&D appreciation courses<br>• Regular customer training programs<br>• Top management customer seminars<br>• In-house customer training programs |

**TABLE 8.6**

Functions of Market Administration, Investment Analysis, and Information Management

| | |
|---|---|
| **Marketing Administration** | • Budgeting<br>• Plan/actual evaluation<br>• Product pricing<br>• Price list management<br>• Discounts policy<br>• Exchange rate policy<br>• Marketing financing<br>• Product price auditing |
| **Investment Analysis** | • New marketing investments<br>• Sales outlet policies<br>• Sales outlet optimization<br>• Dealer vs. branch evaluation<br>• Licensing vs. expansion studies<br>• Handling of exclusives |
| **Information Management** | • Competitive intelligence<br>• Communications with sales force<br>• Corporate newsletter<br>• Customer communications<br>• Information on prices, discounts, and promotions<br>• Reports with press and the media |

describe. Consumer goods companies are in the process of creating in-house testing units that mock up supermarkets from point of sale advertisements to shelf positioning.

Critics say that these are short-term rather than longer-term approaches, limited to some merchandising chores. There are, however, neuroscientists who believe the future for their branch of specialization lies beyond products. They see it spreading into the financial sector, to understand how trust is built and broken down for the banks.

This discussion on wrong way market research and recent effort to reinvent means to measure public opinion is highly related because one of the important longer-range marketing function is promotion campaigns, coordinated with new products coming out of R&D. IBM has been a master of that marketing policy at mainframe times.

In the 1960s and 1970s mainframes were rented rather than sold, but when a new model was about to be introduced IBM salesmen were busy making their customers "offers they could not refuse" to buy the old machines that in a short period of time would have been obsolete. This policy came under the jargon *exclusivity incentives*. Other forms of exclusivity incentives are as follows:

- Joint marketing promotions
- Advertising subsidy in exchange for a commitment not to deal in the goods of a competitor

The exclusivity deal may also take the form of a market share discount that increases when the purchaser gives the manufacturer a greater share of its business. This tends to drive the weaker competitors out of the market. Market discounts may as well take the form of *slotting fees*—marketing stratagem widespread in the food and beverage industry, also used to purchase exclusive shelf space at supermarkets.

Other offers vendors do to their customers within the realm of intensive sales promotion are *bundled rebates*; they come into effect when they purchase more than one product from the vendor's product line. If one of the goods in the rebate bundle is in exceptionally high demand, this strategy is tantamount to *tying*—a practice where a manufacturer makes the sale of a product in high demand conditional to the purchase of a second product that may not be as popular. In case the two products are connected to one another and they are priced together, this is known as *bundling*.

## 8.5 CASE STUDY ON GLOBAL MARKETING BY A MULTINATIONAL COMPANY

Functional lists—like those presented by Tables 8.3–8.6—help in abstracting from organizational charts concentrate on marketing services that must be supported. Because in many cases marketing services overlap with one another, senior management should carefully weed out overlaps and contradictory conditions in order to develop competent business and marketing plans. Tasks requiring plenty of imagination are those of developing sure ways for

- gathering market intelligence
- analyzing and evaluating pricing strategies
- swamping the costs of distribution
- elaborating plans for using agents, dealers, and manufacturers' representatives, and
- handling public relations, advertising, sales promotions, and trade exhibitions

In one of the companies for which I have been consultant to the board, we instituted a new marketing management responsible for world trade, as a separate entity from domestic operations, under the authority of the general manager responsible for the International Group. The new director of international marketing was given authority over

1. Determining the basic marketing objectives and policies on a multinational basis

This required a critical evaluation and definition not of one but of a family of global marketing concepts and their influence on the functions of marketing management per country of operation. To the new global marketing boss reported the chief marketing executives of each foreign subsidiary. However, decision on sales and product policies in a given country, as well as the sales quotas, had to be approved by that country's general manager.

2. Identifying and understanding the company's worldwide markets

This responsibility started with determining all the foreign markets in which the company operates (and not just those where it manufactures).

Part of the mission has been identifying the potential of the market, establishing the characteristics of the company's local marketing effort (buying motives, other influences and habits), uncovering opportunities for greater market penetration, and contributing to planning for new products.

3. Managing market orientation and the product line

This involved a strong influence in determining the composition of the product line by market, establishing the role of market research (Section 8.4), following up on the obtained results, analyzing and evaluating the existing product lines, identifying strengths and weaknesses, bringing to the technical department specific requests for improving or revamping existing products, finding new product ideas, suggesting the nonperforming elimination of products, balancing the product line, and introducing new products to the market.

4. Developing the market plan by control of operations and for the global business

Among the requirements of this mission have been defining the company's major objectives—product line, market position, sales volume, prices and profits, channels of distribution, and so on; forecasting annual sales; developing the marketing plan and supporting plans—sales advertising, sales promotion, product service, and the like; combining individual plans into a coordinated international marketing plan; submitting it to top management for approval; and, after approval, distributing the quantitative goals (quotas) by country.

5. Evaluating international channels of distribution for the company's products

For example, determining the appropriate channels; deciding on new investments in marketing (foreign operations); managing dealerships (where they still exist) and selecting distribution agents and franchised distributors; keeping the distribution relationships profitable to the company; and evaluating the opportunity of launching marketing and sales operations by company subsidiaries rather than distributorships.

6. Building the international sales program

That is, determining the most productive potential to secure sales performance; budgeting the sales functions; determining the number of salesmen needed; evaluating and upgrading sales support; allocating funds to the marketing and sales programs; providing effective coordination of selling activities; continually evaluating the selling program on an international basis; and assuring that this basis is adaptable to each country's specific factors.

7. Conducting international advertising and sales promotions

This responsibility covered the domain of planning the international advertising and sales promotion program, studying alternatives in advertising media and sales promotion methods; selling the program to marketing management by country, and controlling the performance of such a program country by country as well as on a global basis.

8. Providing the subsidiaries with assistance for effective management of their sales force

More specifically, analyzing the salesman's job, reviewing job descriptions and job qualifications, helping to train new and experienced salesmen, reviewing salesmen's compensation and incentives, supervising and developing standards of performance, and conducting performance reviews in connection to international marketing.

Given the aforementioned reasons, an essential function of the multinational staff at headquarters should be to act as a clearing house for capital, skill, and experience. The input received in two global projects I conducted for the American Management Association (AMA), and those that have followed them, suggests that

- instead of trying to control every detail of worldwide operations, the multinational company should set objectives and base evaluations on performance
- well-stated objectives and relatively free reign are important factors in keeping open effective lines of communications. They also affect the acquisition and retention of capable executives and technologists.

- multinational companies should invite foreign subsidiary managers to attend courses and planned meetings at headquarters. These can last for several days so that managers can meet their counterparts from around the world and develop a common base of thought, principles, and aims.
- to increase the sensitivity of headquarters to foreign conditions, managerial positions at the international division should be filled with executives who have experience abroad. A substantial number of these should be foreign nationals.
- acquiring broader knowledge of the situation in which foreign subsidiary operates is certainly a very good step toward better communications and more profitable financial results.

To realize the advantages of a longer-range marketing and sales effort, the International Operations division adopted the *group concept*. This policy is used by several multinational and highly diversified firms as it allows them to decentralize certain functions to the product groups while keeping central control. Typically these product groups vary in structure and market orientation. Activities falling under this perspective include

- Market share
- Sales growth
- Profit margin
- Return on investment

Other duties and efforts of international marketing were to develop corporate prestige nationally and internationally in concert with corporate objectives; search out and enhance product and business opportunities for the company to help offset fluctuations and limitations associated with existing goods; and assure longer-term growth by keeping products and product lines dynamic in coordination with product planning and R&D.

The executives in charge of the International Group appreciated that to fulfill in an able manner its duties in global operations, they needed an experienced body of international specialists at headquarters; business intelligence units staffed for complete, accurate, and analytical service in certain overseas centers; proper market research functions at a number of points of business concentration; collaboration with economic and financial research institutes; and a close working relationship with leading local banks in every commercially important foreign city where the company operates.

## 8.6 CHALLENGES OF A GLOBAL MARKETING STRATEGY

Prior to World War II, American, British, Dutch, French, and German industrial companies were the typical international firms. For some of them, global investments dated back to the golden years of the nineteenth century; British firms, in particular, were instrumental in the creation of an impressive industrial base in the North American continent. After World War II, and especially in the late 1950s and through the 1960s, American companies took the lead, creating the concept of the *multinational firm.*

Typically, in a multinational firm (whether industrial or financial), a substantial portion of the assets and revenues as well as liabilities, relate to its international operations. Two other issues distinguishing a multinational firm are management and marketing policies. Among such a company's goals are the creation and sustenance of a multinational staff in technology and management. At least theoretically, such an enterprise

- may have its headquarters in any nation, though usually they are located in its country of origin
- is characterized by multinational industrial and/or financial operations, but not necessarily by multinational ownership
- taps intelligent resources internationally all the way from research, to manufacturing, finance, and marketing
- deploys resources in several countries from R&D to manufacturing, over and above the marketing/sales operations, and
- exports both from the country of origin and from foreign plants to the markets in which it operates (including the country of origin)

Effectiveness depends on years of experience in the multinational business, a system to enforce accountability, and a strategic marketing plan able to cover every international commitment. Marketing policy must be enforced through steady watch, and marketing management must carefully investigate not only market potential but also

- political climate
- legal requirements
- cost structure and cost evolution
- existing and potential competition, and
- edge provided by the company's brand of products and services

Among the better managed companies, the motives that influenced global expansion emerged from an attitude of management's concern with the future. Decision reflected the study and evaluation of general economics; a market (or natural resources) potential waiting to get tapped; the need to balance business strategy by diversifying activities and profits, and so on.

Although it would be unrealistic not to include in this consideration intuitive moves as well as the impact of personal likes and dislikes, such subjective approaches usually end in trouble. To my experience, there are 12 conditions supporting business confidence, and all of them should be taken into account when examining national markets for expansion:

1. Independence of the judiciary
2. Structure of the legal system
3. Quality and independence of law enforcement industry
4. Laws conforming to those of G-10 markets
5. Political stability
6. Personal and corporate security
7. Absence of stiff money controls
8. Educational system
9. Available skills
10. Labor laws and labor costs
11. Technological infrastructure
12. Other infrastructural supports (water, power, transportation, airports)

The worst possible policy is to give in to crony capitalism, and pass around money under the table asked by (corrupt) local politicians as condition for their support. Crony capitalism has many aspects—from autocratic to oligarchical. In oligarchical, all political power, money, and property is concentrated in a few big families (or companies) that control the mass media, the government budget, labor relations, and international connections.

The autocratic model is a stiff bureaucratic reign. The mandarins at the ministry of finance dictate to the private and public companies what is permissible and what is not. This is done by means of centralized planning, which also maintains illusory exchange rates and interest rates, swamping a company's ability to reach critical restructuring decisions.

Valid and tested methods do exist for analyzing and documenting foreign business opportunities. General management and marketing management can be assisted by an outside firm, or expert international consultant. If a company-wide committee (task force) is set up for restructuring purposes, then such a committee should be headed by a senior executive with global experience and involve people from various functional, marketing, and product lines of expertise.

Each change in market opportunities identified by research and study should be carefully reviewed to measure its relative attractiveness and risks. Such surveys should be compound with the assessment of internal resources and competencies, with a number of questions answered in factual manner:

- Should the company reduce its multinational activities or limit them to certain geographical regions?
- Over what time frame should the firm reevaluate its multinational strategy with a view to revamp it?
- Should management impose some restrictions on the extent of commitments it is willing to consider in a given market in the future?
- What's the economic size below which a project is not worth undertaking, or should it be phased out?

Other questions are related to the acceptable degree of operating flexibility that tends to decrease over time. Is the existing division of foreign markets into concrete, homogeneous categories still valid? Are ongoing changes obliging to alter such division?

- Has the degree of risk associated with "this" or "that" foreign market changed?
- The attitude and policy of the host government toward foreign ventures?

In answering these queries, it is wise to keep in mind that risks in international operations may be political, economic, social, and operational. Political risks may range from minor bureaucratic harassment to outright expropriation. Often difficult to separate from political risks, economic risks may take the form of an unstable currency, galloping inflation, or a policy forbidding the repatriation of profits and/or capital. In some countries, ownership of a foreign venture must be shared with a local interest; this often has major elements of risk.

Political risk and social risks are particularly present in countries with unstable governments and social unrest. They are also pronounced when foreign investments in high-technology industries create fears that the national industries may be deprived of the possibility to compete. Though participation of local interest in a foreign venture decreases the political vulnerability of the enterprise, it results in dilution of control over key managerial decisions.

## 8.7 APPLE, GOOGLE, AND THE POWER OF REGULATORS

Survival in a highly dynamic market requires a culture like that of Sony in its heydays—a company able to quickly innovate, miniaturize, up quality, control cost, and sell through mass marketers. The downturn sets in when corporate culture resembles that of IBM in the late 1980s, where immobilism was the rule. Big Blue was struggling with big-company thinking and its culture made it easier to reject new ideas than to try out new ones judged as dangerous to the installed base.

That is a classic problem. Another problem characterizing modern business and disliked by regulators is the reinvention of cross-holdings by sharing skills at top management level. Several regulators interpret this, and associated exchange of insider information, as unfair competition aimed to hold markets under control.

The building up of a trust is most frequently revealed through managerial and marketing moves. Google and Apple provide a recent example. On August 30, 2006 Eric Schmidt, chief executive of Google, joined Apple's board. He continued being an Apple director for nearly 3 years, eventually quitting under pressure by the Federal Communications Commission (FCC).

Ten months later, in mid-June 2007, Schmidt's membership in Apple's board was interpreted as "a conflict of interest in wait," as Apple encroached into Microsoft's territory by announcing that its Safari web browser would be available to customers of Microsoft's Windows and Vista operating systems. A couple of weeks after that Apple launched its iPhone.

The next move came in November 5, 2007 when Google entered the mobile operating system market with the launch of its Android technology platform. (The software operating system for makers of mobile phones was free of cost.) Then, in mid-February 2008, Google announced that it gets 50 times more searches on its platform from Apple's iPhone than any other mobile handset.

Some analysts cried "foul," and regulators started paying much more attention than before—particularly after mid-2008 when Apple introduced a new version of its mobile phone, the iPhone 3G. It also unveiled a new suite of applications made for the second generation phone on the App Store including

- games,
- auction software, and
- a music-creation tool

Events did not stop there. On September 1, 2008 Google released its own Internet browser Google Chrome, in direct competition to Microsoft's Internet Explorer. Before that month was over, Google entered into smartphone market with the launch of Android powered handset, made by Taiwan-based HTC.

A few months later, in early June 2009, Apple cut the price of the existing 3G phone while launching a faster model, the iPhone 3GS. Within a month Google launched its own PC operating system based on its Chrome browser, to compete with Windows but also with Apple's wares. Rumor had it that the FCC decided to move. On August 3, 2009 Erich Schmidt quit Apple board.

An opinion heard at Silicon Valley is that the FCC's decision seems to have been taken after Google unveiled plans for a PC operating system. The day the market overlap between Google and Apple had significantly widened, was shown through Apple's decision on July 29, 2009 to bank iPhone applications based on Google Voice.

Another reason behind the FCC's decision not to remain a silent observer might have been the fact that smartphones' market grows by leaps and bounds. The way a mid-May 2010 forecast had it, in emerging economies smartphones will reach the mass market as soon as their prices fall below $100.*

All counted, the most basic problem that confronted Apple, Google, and the U.S. regulators has been that modern technologies not only tend to be global by nature (and by name) but also crossover formerly distinct markets, products, and product lines. For instance, in the short span of a dozen years, from the late 1990s to 2010, mobile telephony has become as essential to people as breakfast.

The word "handy"† highlights the importance of a functionality that increased exponentially and by so doing de facto expanded the domain of regulatory authorities. Moreover, at a fast growing pace, this "handy"

---

* *Financial Times*, May 17, 2010.

† Or cellular, or portable or as you like to call it.

became a "computer" not just an Internet gateway with a browser. It is also the tip of the iceberg of a global mobile culture.

The change in administration in Washington might well have played a role in FCC's move. As both Apple and Google continued building their offerings, it was inevitable that the overlap in market offering would grow. The scales were tipped by both parties' mobile devices connected to the web.

There is as well in Silicon Valley a contrarian opinion based on marketing reasons. According to this hypothesis, on the consumer level, the informal Apple–Google partnership combined the best alternative hardware and operating system, with the best platform for additional Internet functions.

Be it what may, public calls for Schmidt's resignation from Apple's board that came with Android grew much louder with the operating system announcement. The Justice Department, too, vowed to step up its scrutiny of dominant companies in technology after long years in which the Bush administration turned a blind eye.

Even if the origin of their business has been different, Apple and Google (as well as Microsoft and others) have crossed into the other party's market. In the aftermath, not only their marketing policies but also the regulators had to account for that. The integration of formerly distinct market domains is exemplified (if not altogether driven) by the iPhone and other smart handsets, which allow users to gain access to the Internet and download mobile applications, including

- productivity tools
- social networking programs
- games, and much else besides

"In the past we tended to run the company to occupy space," said a decade and a half ago, in February 1995, the then chairman and CEO of Citigroup, John S. Reed. In that meeting that he held with financial analysts, Reed vowed that henceforth, "We are going to run the company for performance and not try to be in every market." The merger of handys, computers, and Internet browsers may well allow banks and plenty of other companies to be "too close" to their clients in practically every market—and for this there exists no global regulations.

# 9

# *The Market's Conquest*

## 9.1 THE ANNUAL MARKETING PLAN

The impact of the marketing plan will be proportional to the effort invested in its construction, as well as of the position top management has given to marketing operations. The priority placed on marketing efforts must be examined both in absolute numbers and within the total corporate strategy and structure, which means in conjunction with the other vital functions of the organization.

Typically, the marketing plan comes before establishing sales targets since it determines what, where, to whom, and how to sell which products. The longer-term goal of the marketing effort is that of developing an effective master plan for the market's conquest. This is the top pillar of the structure in Figure 9.1. The other two pillars are

- *products*, including those in the pipeline, and
- *markets*, where the company already operates, and those it plans to enter

Taken together, these three pillars support and animate the sales force, which essentially means the company's foot soldiers and their officers. The sales staff has to be properly selected, trained, informed, and managed within the roadmap established by the marketing plan.

Marketing monitors sales and salesmen results through the establishment of a competent quantitative business plan. This is the "road map" for reaching company goals, a sort of company charter defining—from a marketing perspective—the business we are in, the sales goals to be reached, and share of the market we wish to attain.

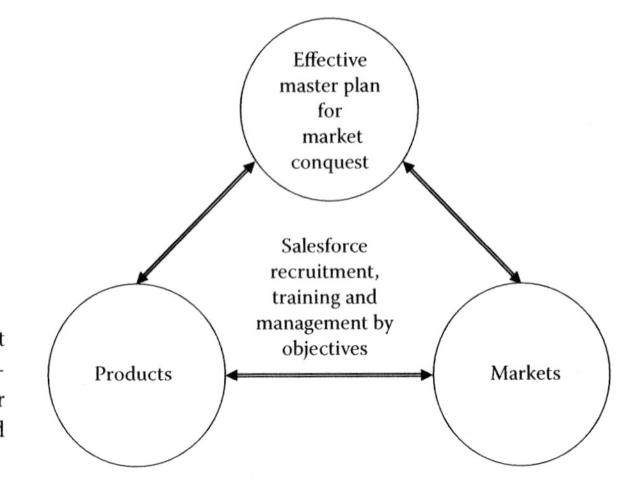


**FIGURE 9.1**
Marketing management starts with an effective plan accounting for products, markets, and the salesforce.

This requires organization and structure. Within the overall definition of corporate marketing functions, different organizational possibilities should be examined depending on the company, its products, and its market(s). The marketing organization can be cast by

- function
- product
- geography
- customer base

or any combination of these factors. The choice should be conditioned by the ability to establish in the best possible way the aforementioned marketing master plan. Typically, this should come after the business plan; lend itself to project planning techniques; and emphasize activities necessary to coordinate with other company functions in defining objectives, schedules, and budgets.

The making of a marketing master plan must involve a careful look at the products, stressing flexibility, emphasizing customer service, and justifying costs. It should identify types of customers: domestic, international, government, private; establish proposal requirements; project marketing contacts; maintain sales discipline; outline the nodes in marketing intelligence gathering; and streamline the reporting system through a timely and accurate flow of marketing information.

To fulfill those aims, the marketing master plan must start with the detailed knowledge of the customers. For this purpose, it will involve market research, product planning, and a pricing strategy, which are behind any marketing decision. This planning must be kept dynamic in its short-, middle- and long-range premises. Key questions to be answered include:

- What does the customer need?
- How might these needs be met?
- What is the size of the market?
- Is the contemplated market there today, or is it 10 years away?
- How much will people pay?
- How profitable will the product be?

Historical data help. Analyses of the total market in terms of current customers, salesmen expertise, products to be sold, annual sales, potential customers stability, multinational perspectives, and competition are some of the marketing plan's prerequisites. Studies need to be more fundamental when planning to enter a new market.

- Who is already making or selling this product?
- Are international companies specializing in this market?
- Is competition broad or narrow based?
- Are potential competitors "big" or "small"?
- What is the market share of each?
- Which may be *our* company's position in the market?
- What's the entry price to reach it?
- What are the strategic considerations that should enter the marketing plan?

Crucial differences exist in an overseas marketing analysis as contrasted to a domestic one. Among the special factors are geography, culture, politics, business style, degree of industrialization, special legislation, export/import restrictions, and skills requirements. The importance of foreign language ability and personality adjustments must be brought into perspective, and also the skill in handling special trade documentation, insurance, trade regulations, available financing, and so on. In short

- plenty of elements enter the marketing plan and influence its mechanism, and
- as they change, they bring up the need for revising planning premises, and often policies and procedures

As these references document, the annual marketing plan is far from being a monolithic business. It must be constructed by product and product line for each of the profit centers (product groups) by both unit and dollar volumes. The best approach is direct. Prior to the annual profit planning meeting, marketing should submit its plans and objectives for the coming year by profit center, product line, and single product. Idem for markets.

Medium- and longer-range marketing plans should also be presented by product line, but not necessarily by profit center. All marketing plans begin with a forecast of the total market over an established time range, for instance, for 1-, 5-, and 10-year periods based on a report by the market research and sales forecasting section.

In the execution of his or her duties, the director of marketing should be aided by field (sales) executives. However, he should be himself responsible for carefully studying, (altering if need be) and approving implementation plans to meet the company's objectives with regard to sales volume by product line, profit margins, and other criteria.

If there is no major change in conditions beyond the company's control, after the yearly objectives have been established they should remain in effect to serve as a benchmark for measuring results. The 5- and 10-year forecasts, and plans based on them, should be reviewed annually (or more frequently) and revised when necessary.

Long-range plans will largely reflect anticipated levels for the general economy with allowances for various economic conditions. If the economy operates at a substantially higher or lower level than anticipated, the objectives may become unrealistic and should be recast.

The recruitment and training of salespersons, tooling of a service (maintenance) engineering network, and preparation of a customer training capability are also integral parts of marketing planning. They demand larger capital outlays than corresponding production requirements as well as considerable lead time.

## 9.2 MAKING THE MARKETING PLAN

All planning programs require objectives to work with. An objective should be achievable within the contemplated planning effort and within the limits of reality. Realistic objectives have a direct relation to the problem to be solved; do not exceed, in terms of demands, the company's resources needed to know commitments involved; being made; and are able to assure that these commitments are in line with the firm's strategic plan and ethical values.

The sense of these references is best explained through a saying, which states that "a serious person not only keeps the promises which he makes, but also makes the promises that he can keep." This applies hand-in-glove to both people and organizations, because organizations are made of people.

The objectives used in making a marketing plan should be noncontradictory to other objectives that are already established. They should generally fit within the structure and mission of the organization; be consistent, measurable, and tangible; and supply the structure needed to make meaningful decisions, rather than leaving ambiguities about the chosen direction and/or alternative paths.

It is self-evident that simply committing sales objectives to paper does not constitute a marketing plan. Objectives must be so phrased as to give guidance and generate ideas, but plans do not crystallize until the decisions on *commitments* are done. Under this perspective four basic issues affect the creation and execution of a marketing plan:

1. Formulation of the task

More precisely, defining the scope, examining alternative methods of procuring answers, selecting a solution based on cost/benefit and resource availability, and specifying the action required to develop the results needed to carry the day in the marketing and sales effort.

2. Development of necessary information

This requires assembling the facts; postulating what is required to confront uncertainties (connected to the chosen course and its alternatives); describing these alternatives in a comprehensive, quantitative form, and establishing marketing plans based on them, which—following step No. 3—can be kept in reserve (Plan B, Plan C).

3. Evaluation of alternatives and decisions on action

The best way for making factual choices is to convert alternatives to terms that can be compared; establish criteria for making a selection; apply these criteria; compare alternatives in terms of cost/effectiveness, and examine their feasibility and desirability. Subsequent to this decide on the course of action to be followed.

4. Translation of decisions into executive statements

The EVP marketing must clearly identify each action; account for resources involved; outline consequences expected and their timing; and provide for necessary control for the interim measurement of progress. This must be phrased in both qualitative and quantitative terms, so that corrective steps can be taken when it is still time to do so.

After the company is committed to a course of action success rests on the formality and observance of the planning program. This runs contrary to the practice of many firms and CEOs that make the mistake of thinking that leaving procedural issues undefined supposedly to "guarantee more options" for marketing management. Nothing is further from the truth. The only thing guaranteed by undefined or inadequately defined responsibilities and missions is confusion.

In addition, positioning *our* company's marketing effort is not just a matter of making some choices about markets, products, market share, and the like. These decisions have to be backed up, to a very significant extent, by financial and technical data. Among them

- *marketing costs*, including stable establishment(s) and advertising
- *sales costs*, particularly if salesmen are primarily on salaries or on, commissions
- *distribution costs*, which imply a technical organization beyond sales proper (for instance, including field maintenance)
- *R&D costs* to promote the company's image in the local market and gains from local knowledge and skill
- *quality standards*, which must be higher than those observed among competitors, and
- *legal protection costs*. Legal costs should never be underestimated. All industries today (most particularly the financial) are open to class and other actions

The importance of marketing-related tasks, and skills that they require, are best appreciated by the diversity of the jobs entering into the marketing master plan. The cases brought to the reader's attention as well as product life cycle studies and overall marketing expenditures (justifying and financing a rigorous marketing effort), are but a few examples.

Product life cycle studies are integral part of marketing planning. They must focus on all phases of product life, from R&D to manufacturing, product sales, and product field services. Pricing and cost analyses associated to product through their life cycles are a "must," and this is also true of evaluating customer feedbacks, reports from field engineering, and bottlenecks in distribution and logistics.

Other large area of marketing endeavors concerns costs and benefits. In a study with a business equipment company I was consultant we established that the marketing expenditures necessary to launch a new personal computer line (a "first effort" in this field by the subject firm) exceeded the corresponding engineering and production investment by a ratio of 3:1. For another business equipment manufacturer, this ratio was 2.2:1. In many cases, indeed

- the marketing effort is more capital-hungry than engineering, and
- marketing expenditures have to be repeated time and again as the characteristics of products change.

The intensity of a company's need for capital to finance the marketing end of its expanding operations both influences and is influenced by the marketing plan. This intensity will be based on senior management decisions and their implementation, as well as industrial and scientific developments, shifts and tendencies in the economy, as well as market liquidity and money rates.

Solutions are in no way linear between a company's creditworthiness and sources of financing. A company's appeal for funds to banks and capital markets changes with time. A company's availability to get the funds it needs and the cost of money change with its credit rating. In the larger number of cases and, therefore, in the typical instance, the salient problem in creditworthiness has much to do with the product line and with marketing plan.

The intensity of marketing expenditures is much lower in a situation characterized by a stable nontechnological product line, with well-established marketing channels. In these cases production expenditures may well exceed those of marketing. When marketing uses such interfaces as franchised dealers and agents, the associated investment of the firm largely consists of selection, structuring, and subsequent control of dealerships, but not of the actual establishment and operation, which characterizes the firm's own distribution channels.

Other objectives than those treated in the preceding paragraphs concern the dimensioning of necessary human, physical, and financial resources, as well the observance of lead times needed for developing new products, new markets, new business opportunities, and better competitive positions. A most valuable contribution is made by thinking out of the box, as Section 9.3 documents.

In conclusion, a marketing planning program is instituted to ensure market leadership—therefore growth—and to guarantee that the company gets its share in market evolution. Since most of these functions are measured in years, there must be in place a methodology helping to integrate plans of short, medium, and longer ranges into one marketing master plan. This is a demanding task requiring a great deal of skill, experience, and foresight.

## 9.3 BLOOMBERG FINANCIAL MARKETS: A CASE STUDY

*Bloomberg Financial Markets*, the securities trading computerized information system, provides a first-class example on how successful a marketing plan can be *if* those making it are able to use both their experience and their imagination. Bloomberg distributes information on domestic and foreign bonds, stocks, futures, options, and other publicly traded securities. It also features interesting interviews and acts as a supplier of news through *Bloomberg Business News*.

Today, the financial news market is dominated by Bloomberg, whose terminals come with analytical tools that have been the company's spearhead in conquering the market of banks, insurance companies, other financial institutions, and corporate treasuries. Reuters, which had a practical monopoly on such terminals in the 1980s and first half of the 1990s, fell increasingly behind in providing what professionals needed to do their daily work.

When he set up his company, Michael Bloomberg saw right on which way the chips would fall—and formulated both his product offering and his marketing task accordingly. Finance directors, traders, and analysts want to know not just the price of a bond, but the bond's price history and its performance compared with other investments. This is available on Bloomberg.

As a result of product innovation and a sound marketing plan, Bloomberg gained market share while Reuters' financial information business which, at the time, accounted for two-thirds of Reuters' revenues, went flat. Furthermore, the incumbent made a disproportionate share of its money in Europe, thereby opening its flank in the American market.

Like IBM in the 1980s, Reuters' basic problem was that for years it dominated the wrong side of the business. IBM bet on the hypothesis that the mainframes era will never end. Reuters' false sense of security came from the fact that it was strong on trading floors, but as investment banks merged, the number of traders, and therefore the number of screens, has been falling.

- Reuters' marketing plan was one-sided and therefore wanting, damaging by so much the incumbent's defenses.
- By contrast, Bloomberg's marketing plan was novel, capitalizing on the fact that by making a couple of big mistakes Reuters might lose its market.

Michael Bloomberg's marketing hypothesis was that the volatility in interest rates and currencies—which followed the Smithsonian Agreement, deregulation, and globalization—changed the rules of the game. Analysis, not just information, was becoming more important than ever.

Quick access to the latest financial data, price blips, or news events continued being essential to survival for people and companies who move billions of dollars around the world. But increasingly end-users required more than that.

- Barebone information on financials used to be the exclusive domain of Reuters Holdings, ADP, Quotron, and Telerate.
- But Michael Bloomberg came up from under and took over market leadership by offering *added value*, because he was an alert entrepreneur whereas his competitors were parts of vast bureaucracies.

True enough, some of the minor incumbents also detected the new marketing trend. Quotron moved in the same direction and it had an early lead, but became a disaster under Citicorp. Telerate, for which Dow Jones paid $1.6 billion, was slow to upgrade its technology to keep from losing its customers. (That mistake is repeated by many Western companies today, particularly in Europe, who have fallen in love with their legacy information technology.)

Personalities did play a rather significant role. Michael Bloomberg was a trader, not an information scientist, and he had the salt of the earth. He joined Salomon Brothers and went on to become head of equity trading and sales for 5 years. His engineering and financial background helped.

- As an engineer, he understood how computer systems work.
- As a trader he knew how to use them for successful dealing.

After the merger of Salomon with Phibro, Bloomberg moved to Merrill Lynch. When he asked for a system similar to the one he himself later established, he was told by the EDPers* to wait 6 months for a feasibility study. They needed that long just to examine the practicability of his request. Such slow deliverables are commonplace in IT—but not for Michael Bloomberg who quit to make his own company and provide precisely the kind of service he asked for.

Bloomberg Financial Markets left its competitors in the dust because its promoter cared about the competitiveness of *his clients*. This has shown that he was both a strategic planner and first-class marketing executive. Since 1991 Bloomberg also anticipated the day when there will not be room for five or six market quotations and news services. Many users, he predicted,

- will pare down to one or two machines, and
- they will choose only those that offer everything needed at their fingertips

It follows logically from this reasoning that the survivor must pack his terminals with every kind of information imaginable, as well as analytics.

Indeed, early on Bloomberg went beyond financial quotes and news adding weather forecasts, airlines schedules, real estate listings, classified advertisements for jobs, bank loans, and comprehensive access to its databases—changing the former financial terminals to multimedia machines.

- By adopting an open architecture, Bloomberg Financial Markets expanded its reach.
- By being ahead of the curve, the firm took the lead in innovation, which is precisely what makes a successful enterprise whether in finance or anywhere else.

---

* Information technologists of Paleolithic "electronic data processing" thinking.

This is one of the case studies which document that marketing policies followed by companies that have been successful in their line of activity are in no way cookbook solutions. Moreover, management decisions and planning commitments must be translated into executive statements capable of being put into action; with actions controlled to ensure their conformity to plans.

As the case we have just examined demonstrates, plans must be kept dynamic. Therefore they should be made in a way that they are easily revised. This is a full cycle, to be supported through the marketing planning program by means of clear directives and well-established steps.

In fact, the word *program* might even be a misnomer. The object is not only programming as such, but also the thinking and organizing that go into it—which is typically longer term. Marketing management should never be satisfied with short-range results but, instead, position the company to match the market's movement and face the future.

A crucial part of marketing planning is to account for adversity before it hits like a hammer. Errors made with logistics provide an example. In the aftermath of the first Persian Gulf war, Desert Storm looked like a triumph of logistics. Moving 7 million tons of supplies for 550,000 troops in short order was a feat. Only later did little-noticed Defense Department reviews reveal that the 1991 operation was flawed:

- Half the 40,000 containers shipped to the desert by the U.S. Army (including $2.7 billion worth of spare parts) went unused.
- This spoilage happened because soldiers did not know what was in many of the containers, or where they had to be channeled.

Learning from this experience, in 1995 the Pentagon set new goals to replace the just-in-case delivery mentality with a streamlined just-in-time system based on business premises. The objective has been to shrink inventories and speed deliveries—which is also a "must" in every industrial firm. The right planning is very important, but never forget the issues associated to the execution.

## 9.4 MARKETING THE USE OF REVERSE INNOVATION

Research and development is risk capital. So is the manufacturing engineering expense, as well as the marketing activities, of the sales effort—the development and control of the client base. With these premises in mind R&D, manufacturing engineering, and marketing form a unified total that should target not only new products but also *reverse innovation.*

Positioning our organization in a way to capitalize upon marketing ingenuity is a polyvalent enterprise seeking to be ahead of market drives, outpace competitors, and promote profitability. Successful marketing management considerations can be better understood if they are specific and well focused. Among other prerequisites this involves the interaction of three factors:

- developing a business opportunity analysis,
- evaluating alternative product and service scenarios, and
- translating decisions into specific marketing plans

There is no surprise that with globalization, marketing and research and development costs correlate more than ever before—or that both activities have been internationalized. GE's health-care laboratory in Bangalore* contains some of the company's most sophisticated products. This

- promotes GE's marketing image in India, and
- it assumes that the company benefits from local R&D skills

For instance, a device that came out of GE's Bangalore laboratory is a hand-held electrocardiogram (ECG) called the Mac 400. It is a device small enough to fit into a backpack and can run on batteries as well as on electric outlets. In addition, it sells for $800, instead of $2,000 for a conventional ECG, reducing the cost of an ECG test to just $1 per patient.

- Mac 400 is part of a process known as *reverse innovation.*
- The concept is brilliant, featuring simplification and low cost at its core.

Instead of adding ever more sophisticated gear and complexity, reverse innovation strips products down to their bare essentials while retaining their functionality. Marketing both provides input to and gains mileage from reverse innovation because this process is not just about redesigning

* General Electric's health-care arm spent more than $50 million in the past few years to build a major R&D center in Bangalore.

products. It also involves rethinking entire production processes and marketing models:

- Squeezing costs so sophisticated products can reach more customers, and
- Accepting thinner profit margins to gain volume, provided the marketing effort is successful

As this GE example documents, it is not without reason that global firms are increasing their R&D investments in emerging markets—and reverse innovation is one of the keys in understanding the trend. Indeed, American companies of the Fortune 500 have 98 R&D facilities in China and 63 in India.

Neither is reverse innovation a notion that came out of the blue. In the 1960s and 1970s the Japanese redefined marketing goals by inventing *lean manufacturing*. Emerging countries today are going way beyond that by turning to the research, development, and engineering roots of product design.

- They are altering the rules of the global game down to its very basics.
- They are turning on their head marketing methods, which are based on classical production and distribution concepts.

The fact is that marketing efforts have to be totally revamped to face the challenges described in these bullets. *If* Western companies don't do it, *then* eastern companies will do it for them. The United Nations World Investment Report calculates that there are now around 21,500 multinationals based in emerging countries—most of them in Asia.

A different way of looking at this issue is that the time for complacency is past. Revamped product design, manufacturing, and marketing policies and methods should see to it that a serious solid strategic plan provides the needed mechanisms for appraisal of prevailing procedures and rapid corrective action.

The use of feedback, discussed in Chapter 7 in connection to management control, is integral part of properly positioning the marketing effort because it constitutes a program of surveillance. As such, it helps to identify and size-up changes in the business environment, contributing to the sense of specific product opportunities and threats implicit in ongoing changes.

Internal control can be instrumental in empowering planning reviews serving to sharpen management judgment as to whether a new mission, competitive strategy, or program of action is called for. This way, new competitive marketing plans can be established with valid goals as targets for company performance.

Steady reviews of plan versus actual, therefore ongoing achievements, make the marketing planning process continuous and self-adjusting. It also underlines the need that, after marketing policies and specific goals are selected, intermediary steps are set up in the form of shorter-range plans—with review processes employed to gauge such steps.

- Such reviews involve the measurement of performance.
- Through them, facts are analyzed and judged so that conclusions can be distilled from the output of operations.

It would be a mistake to think that inverse innovation calls for less marketing effort. If anything, it may call for more effort. Marketing plans should look ahead and determine what must be done to reposition the company's sales force, while reviews look back, establishing what actually has happened. There is an intimate interplay between planning and reviews.

The share of a market repositioned through inverse innovation, sales growth, profit margin, type of investments, and so on are quantitatively expressed; hence, amenable to objective evaluations. As indicators, they suggest the degree of stability of the company, whether it has been expanding or contracting, its influence within a given market, and where strengths and weaknesses lie.

One of the unwanted consequences of inverse innovation that must be confronted head-on, is that past marketing objectives and policies can quickly become obsolete. *If* the enterprise does not alter its course to meet the changing standards of price downsizing, *then* it will lose customers, personnel, prestige, and profit margins.

It is as well self-evident that the persons most interested in correcting undesirable marketing conditions should be given the information support necessary to improve results. Conversely, unusually good performance should not only be the basis for appropriate credit and reward but also be used as a starting point to seek similar results in areas showing a less satisfactory outcome.

In conclusion, as the examples provided in this section document, there is no monopoly of ideas. Improvements in products and methods are often found by thinking out of the box. An alert marketing management should use all available means for improving efficiency at any opportunity. Complacency has no place in the modern enterprise.

## 9.5 CONQUERING THE MARKET THROUGH EMPIRE BUILDING: GENEEN AND CHAMBERS

Contrary to other masters of industry, like Andrew Carnegie, whose business principle was to focus in one product line, Harold Geneen's idea was the conglomerate—and his vehicle ITT.* An accountant by training, Geneen postulated that a company could successfully invest in any sort of business anywhere, and manage this business in a way to be ahead of the curve.

In the 1960s Geneen found himself at the top of a mainly technology firm made up of units AT&T and RCA bought by a tycoon in the 1920s and 1930s. The line of command was twisted, but Geneen imposed discipline on ITT's many and diverse business units by setting strict financial targets. Then,

- he kept on growing by buying time, through acquiring a lot of firms, and
- he did so by using as currency ITT's own highly rated and highly leveraged shares

This has been an incarnation of the concept of market control through financial control originally invented six decades earlier by Dr. J.P. Morgan, with the Trust. Its implementation came naturally to Geneen because of his past experience and his attention to financial detail combined with self-confidence. This propelled him up the ladder of various big American companies.

The chance to satisfy his desire to be his own boss came in 1959. The mission was to run what was then a lackluster collection of mainly telephone companies, which he turned into a conglomerate. His critics lost no opportunity to say that the conglomerate he built also had political overtone. It was a *supranational organization* not accountable to individual governments and supervisory authorities; indeed, trying to subvert them.

By the time Harold Geneen left ITT, he had bought 350 companies in 80 countries, creating an entity with sales of $17 billion, which was big

* Once known as International Telephone and Telegraph.

money at the time. Comparatively speaking, it represented more than 20 times ITT's annual turnover when he became its CEO. His strategy ran on the principle that

- first-class financial management could be applied successfully to any business, and
- sound finances is the precondition and precursor to conquering markets—national and global

Anecdotal evidence suggests that Geneen worked a 70-hour week. Though the transmission of power to his successor was far from being simple and linear, he had a successful retirement as a private investor. ITT, however, fared less well under the rather inefficient management of his successor. After Harold Geneen, ITT fell out of the go-go companies' radar screen.

In its way, Cisco has been the late 1990s ITT, a feat that continued, albeit with less pronounced overtones, during the first decade of this century. When in January 1995, John Chambers was named chief executive of Cisco Systems the company had 3000 employees and sales of approximately $2 billion a year. In a little over half a decade it had become a telecommunications powerhouse, with 30,000 employees and annual revenues nearing $20 billion.*

The company's core business has much to do with this growth. The added value was management, including marketing wizardry as well as a score of acquisitions. Cisco makes and markets the essential plumbing of the Internet: routers, hubs, and switches. Through these products it became one of the larger U.S. companies—indeed the fastest-growing technology firm in the late 1990s.

Chambers was an alumni of IBM and Wang, and part of the added value contributed to Cisco was that of changing the San Jose–based company from being a one product business to an integrated communications products provider. Thousands of people work for Cisco in 54 countries. Chambers also added customer focus with his insistence that

- marketing must be global and at the same time sensitive to local user requirements, and
- no technology should be rejected on "not-invented-here" grounds; technology's role is international

* Notice, however, that the ratio of revenues per employee did not really change over those years.

As these bullets show between the lines, Cisco takes a pragmatic approach to product and market planning. If a useful technology can be more quickly bought than developed by its own engineers, then it is bought. The first acquisition was the purchase in 1993 of Crescendo, an ethernet switching company, for $95 million—and it was prompted by Boeing.

About to lose Boeing's business, Chambers asked what Cisco had to do to keep it. The reply was, "make the new switching technology a core competence, and do so fast." Next to this, Cisco faced technological demands spurred by the convergence of the internet, telephone, and television. Good news and bad news were associated to this convergence:

- It brought new markets,
- but these markets posed unfamiliar requirements.

Cisco was swift in facing the challenge because experience had shaped Chambers' mind, helping him to avoid the arrogance that can make big and successful technology firms remote and unresponsive to their customers (the old IBM's syndrome). It is rumored that to assure that he keeps in touch with his customers, Chambers goes every evening through voice mails from managers dealing with customers on the critical list. The goal is to hear not just their words but also their emotion.

"... We've always made decisions based on what we think is best for our employees and shareholders and the company in the long run and not on short-run gyrations," says Chambers.* To enhance employee loyalty and reduce turnover, Cisco ties compensation programs directly to results identifying customer satisfaction. Here are, in a nutshell, its market-centered management's principles:

- Make your customers the center of your culture.
- Thrive on change and its management.
- Empower every employee to increase productivity and improve retention.
- Build strong partnerships in an industry more innovative than ever.
- Keep two-way communication and trust, required by teamwork.

---

* *Wall Street Journal*, June 2–3, 2000.

Teamwork, however, does not mean nepotism. Anecdotal evidence suggests that Cisco's culture welcomes competition because competitors are the promoters of innovation—and they are forcing a company's management to act quicker. "We will have more market share 3 to 5 years from now because we have good competitors," John Chambers suggests.

Human capital has been a success factor. The firm has seen the importance of keeping good people as a means to improve deliverables and to avoid unnecessary disruptions in the supply chain. What Cisco is generally buying through company acquisitions is not just technology and market share, but young entrepreneurs as well. Chambers style of decentralized management sees to it that people who do well are given a lot of freedom; people who do not, leave. "It's called accountability," says Cisco's CEO.*

## 9.6 GATES AND MICROSOFT'S HOLLYWOOD MARKETING MACHINE

Back in 1979, Ross Perot met with William H. Gates III to discuss buying his fledgling start-up company, Microsoft. The then 23-year-old Gates was asking between $40 million and $60 million. Perot, who at the time was the boss of Electronic Data Systems (EDS), thought the price was much too high. A dozen years down the line Perot told the *Seattle Times* he still kicked himself over that lost opportunity.†

As ultra-brief personal résumé, Bill Gates entered Harvard but dropped out to found Microsoft in 1975, 4 years prior to the Ross Perot meeting. The new company's first product was a version of Basic, the programming language, for the Altair 8800, one of the first PCs. Considered by many as the world's first personal computer, Altair 8800 sold as an assembled marketable unit and it preceded Apple by a few years.

The next thing Gates played with (and eventually won) was an operating system for personal computers. By 1980, after long delay and internal struggles, IBM had decided to build personal computers and needed a PC operating system. While Big Blue employed thousands of capable software builders, these had the bad habit (inherited from mainframes) to take ages to produce deliverables—with all that meant in terms of time to market.

* *The Economist*, March 28, 1999.

† Communications of the ACM, August 1992, Vol. 35, No. 8.

Moreover, the culture of IBM's software professionals was not right for the new epoch. It had been mainframe-based, and PCs were a different ball game. In the aftermath, IBM hired Microsoft to build the operating system for its PC. Microsoft bought Q-DOS from a company called Seattle Computer Products and retailored it for Big Blue's PC.

- The OS was released in August 1981 as disc operating system (DOS).
- It was followed into the market by a huge flock of clones, which saw to it that DOS established itself as a sort of school uniform of personal computing.

By the time Apple released the Macintosh (in January 1984) it was too late to dethrone the DOS, even if compared to Apple's sophisticated operating system, DOS was obsolete.* Politics and personal rivalries, not OS obsolescence, are the reasons why IBM's and Microsoft's partnership oscillated between ups and downs.

When this curious partnership hit the rocks, Big Blue's OS/2 died out along with the PC on which it ran.† By then Microsoft was standing squarely on its feet and did not need anymore of IBM's rich contracts to make ends meet.

Although the design of the first PC language and upgrade of the first PC operating system were not minor feats, the real genius of Bill Gates does not reside there, but rather in his marketing insight. He can easily be credited with two concepts that equal in discovery Moore's Law. That

1. personal computing could be a high-volume, low margin but highly profitable business, and
2. unlike mainframes, minis and maxis, PCs and their software can be sold like hotcakes—but to do so requires a Hollywood setting rather than a gray suit, well-combed salesforce

In regard to issue No. 1, Gates realized that falling hardware costs, combined with the trivial expense of making extra copies of standard software, would turn the computer business on its head. Both functionality and productivity would see to it that personal computers could be

---

* Only by May 1990, Microsoft perfected its own version of Apple windows and called it Microsoft Windows 3.0—a huge marketing (but not necessarily technical) hit.

† IBM's PC division was finally sold to China's Lenovo; more or less a "me too" company.

on every desk and in every home—a vast market justifying slim per unit profit margins.

Profits would come from selling a lot of them cheaply, not from servicing a few organizations at a high price—mainframe style. Gates also correctly saw that profits from making hardware and writing software could be stronger as separate product lines, even if some firms clung on to both the operating system and the hardware—like mainframe companies did.*

Based on these premises and working in unison, in what has been the closest thing to a partnership, Microsoft and Intel blew the then prevailing computers business model apart. Criticized as the Wintel duopoly, their synergy tightly controlled the PC market to the benefit of each and both firms.

But it is not enough for a company to come up with an ingenious idea. To gain market breadth and appreciation, it must also be seen as a front-runner—something that seems to have landed from another planet, or has never been seen before. That's where the Hollywood pizzazz came into the picture.

Timing the launch of the Windows OS at the heart of Comdex 1983 Microsoft's Hollywood-by-Las Vegas event has been a public relation's positioning at its finest. Bill Gates applied all the way his No. 2 dictum, which provided for lots of the company's employees and shareholders (not only for the firm) the opportunity to open a bank by the side. Masterly executed, the Las Vegas 1983 event saw to it that

- a new software star was born, and
- it was positioned at the very top of the personal computers industry's food chain

Credit for the Comdex event's design goes to Rowland Hanson, Bob Lorsch, and Pam Edstrom but the push (and the money) came from Gates. The able hands he hired for that Hollywood extravaganza knew that they were never going to get the company's planned marketing blitz by working through normal channels. Those were all taken, and they were not that brilliant anyway. So they decided to make the impossible happen—and at affordable cost.

Here is how Jennifer Edstrom and Marlin Eller relate the event: "There was not a taxi on the strip not promoting Windows … Stickers were all

* An evident reference to Apple. This principle, however, is no more true.

over the backseats of cabs; the drivers wore Windows buttons ... These same buttons were handed out at the booths of every hardware manufacturer that supported Windows ... People could not go to bed without Windows."*

With the Hollywood-at-Las-Vegas Windows blitz, Microsoft went from being "one of the software companies" to being *the* PC software company—ironically for a product that was still far from being released. Gates became a celebrity, and people got so convinced that Microsoft meant something that they did not even need a reminder like the huge Windows panel right outside the front lobby of the Comdex Convention Center.

In retrospect, the most beautiful part of the Comdex 1983 extravaganza was that Windows, the masterpiece of the show, did not even exist. In 1983 it was a label, with some concepts attached to it. Essentially, it was kept alive as a place holder, and it would not become a marketable PC operating system until many years later attracting the interest of companies in the late 1980s/early 1990s with the Windows NT version.

By the late 1980s/early 1990s, however, Microsoft was *the* PC software firm. Its range of products was expanding. New opportunities queued up. The World Wide Web emerged in 1994, making browsers necessary. Netscape was founded that same year. Microsoft's first browser, Internet Explorer 1.0, was licensed from Spyglass, a software company, and thrust into the market. The rest is history—and part of history is the fact that nearly three decades after Hollywood-by-Las-Vegas Microsoft is no more a go-go company. Its war lords have aged.

## 9.7 MICROSOFT'S MARKETING METHODS: THE EMPIRE STRUCK BACK

In the early years of its life, within less than one-and-a half decades, Microsoft emerged from a start-up to become the most important single force in the software industry. Over the years it did so by steering itself from dominance in operating systems into a commanding position in applications programs and office suits. It is to its credit that it addressed the software needs of the Internet without losing control of the OS; and to its debit that after that came stagnation.

* Jennifer Edstrom and Marlin Eller, *Barbarians Led by Bill Gates*, Henry Holt, New York, 1998.

There have been as well stones thrown at the Microsoft tree. In any industry, a dominant position by one company is thought to be associated with abuse. It also invites greater scrutiny by government—and not only scrutiny (more on this later).

Critics say that Microsoft has been using its power in ways that are just like IBM's. They add that unless *they*, the Justice Department and Attorney Generals from different states, change the way Microsoft licenses its OS and other wares to major PC manufacturers, it will be nearly impossible for smaller firms to compete.

- Microsoft answers that PC makers have been choosing several different ways to license operating systems.
- But makers of applications software allege that Microsoft has advance details of changes in its OS software, and the company is slow to share vital information.

This is a hot argument, and it is not destined to fade away. The more sophisticated is the technology, the greater the chances that leader takes all. Many people doubt Microsoft's response that it freely shares its knowledge with the industry and enjoys no inside advantage in developing applications routines. Regulators don't have much of a chance to sort out these claims. After all, they depend on experts who (in large number) have adopted "this" or "that" position in regard to how they look at the dominance of the software market.

Critics accuse Microsoft of having taken more than one leaf out of IBM's book. For example, to preempt moves by competitors products it sometimes announces products years before they actually exist—the classical strategy of IBM's "broken down cash register." Microsoft responds that it is to its own advantage to let outside software developers know its directions in system software.

Another ongoing argument about market dominance has centered on pricing, and it goes like this: Microsoft can offer low prices in two ways. First, by including extra routines with its OS; second, by using profits from operating system sales to support low pricing of applications programs. Microsoft responds that this is an industry-wide trend; as operating systems evolve, more features are included as integral part of the OS.

The irony of all this is that both sides of the argument are right. In a very dynamic, hypercharged industry where no standards have been established because they are anathema to the players (and to innovation), the field is open to everybody. Market leadership does not necessarily go the company that is swift and leads the market, but this is the way to bet. In addition

- precisely, because this is a hypercompetitive market place, scale is not positive all by itself, and
- smaller companies have an edge when supported by swiftness, innovation, and the ability to turn on a dime—and this typically characterize the small companies not the big ones

Critics say Microsoft is a monopoly. It is an argument that rages for more than a decade. In the Fall Comdex of November 1999, in Las Vegas, an audience of about 500 computer experts was asked, "Is Microsoft a monopoly?" 57 percent answered "yes;" 40 percent said "no;" and 3 percent were uncertain. The result of this vote is quite pertinent, because it came at the climax of the Clinton Administration's effort to blockbuster Microsoft.

The "yes" had it, but when reference was made to the Justice Department's case, one of the persons who said "yes" added, "I hate to see the dismembering of Microsoft." Another person who said "yes" suggested, "If Microsoft stays as one entity its stock price may double at best, but chance of tripling is very low." In other terms, the company will worth more as the sum of its parts.

One of the experts who voted "no" was of the opinion that Microsoft's position is well-earned over many years, and having in the market one de facto standard in OS, helps everybody. Another person commented, "Microsoft's position has been created honestly. Only the losers complain."

As the debate on this issue went on, a sense of the meeting emerged that having a hold of the market goes against American culture. Examples were given of Ford and Rockefeller—the two, however, contradict the thesis that, at least over some time, a monopoly is just one big negative factor.

The dominant position of the Ford Motor Company was overtaken in the late 1920s by GM, which came up from under. As for Standard Oil, it was broken up in application of the antitrust law but a couple of decades later each one of the companies that came out of it has been bigger than the original Standard Oil ever was.

An important comment heard during the same conference, both linking and contrasting Bill Gates to Henry Ford and John D. Rockefeller, was that they invited upon themselves the government's enmity. This was mellowed by the argument that "all these people created rich foundations. Gates should do the same." He did (The Bill and Melinda Gates Foundation is today the richest in the world. In the early years of this century, it has a huge endowment).*

The Clinton administration started a huge probe into Microsoft in 1996. DOJ alleged anticompetitive practices. But the way an investors report put it, "The Microsoft case has always been the Federal government's version of a protection racket."†

In other terms, the politicians behind the probe *wanted money*—which in a first time Gates refused to pay. Let's follow a little closer that line of reasoning. The idea behind what Mark L. Melcher and Stephen R. Soukup wrote in a financial analysis report by Prudential Securities, was that Bill Gates had had the audacity to try to function in the modern bureaucratic state without paying the requisite tribute to the political capos in Washington. Based on that they suggested that antitrust chief Joel Klein and the crowd over at Justice had decided to use Gates as an example for other technology entrepreneurs who might try to function independently.

> Gates was the guy at the bottom of the lake in the cement shoes; the guy who didn't pay up.‡

What Melcher and Soukup particularly lamented is that this antitrust case really centered on an industry that once avoided the inefficiencies of the bureaucratic state. It also did what it could to escape most of the regulatory constraints that hinder traditional, old-economy businesses. To the opinion of the aforementioned Prudential Securities analysts

- Gates paid little or none of the tribute demanded by the bureaucratic/political machine, and
- he gave little to either party, and did not have to spend what IBM, for instance, was spending on *political overhead*

The empire, however, struck back and in early June 2000 Microsoft was ordered by a U.S. judge to split itself up to restore competition in the

---

* D. N. Chorafas, *The Management of Philanthropy in the 21st Century*, Institutional Investor, New York, 2002.

† Prudential Securities, Investor Weekly, April 2000.

‡ Idem.

software industry. The company appealed the court ruling, but the bureaucrats counterattack spread across the Atlantic. The European Union did not fail to follow the Justice Department's lead—probably for the same reasons of paying the dues to political overhead.

What the reader should retain is that the sweep and substance of government rulings against profitable firms create historical precedence that handicaps future growth by other smaller but innovative firms. "When you have a global marketplace, the results of a breakup are much less predictable and are likely to reverberate around the world," said Brad Smith, Microsoft general counsel.*

With hindsight Smith might have added that the government should take stock of the AT&T experience. Followed by deep mismanagement of the remains, the dismembering of Ma Bell killed the formerly proud firm.† Neither was the argument that Microsoft was killing innovation any good.

Since it became a big company Microsoft was no more the innovator it used to be; internal bureaucracy slowed down the growing of ideas. At the same time, it did not have any way to stop others from being creative—and these "others" had learned some lessons in management and in marketing by watching IBM falter and Digital Equipment crash and burn.

---

* *International Herald Tribune*, June 9, 2000.

† The best quotation regarding the split into different firms comes from Microsoft Europe's chairman, Michel Lacombe. He said that "our assets are people, and it's not a simple thing to break up people. All the synergies would be lost." And most likely not only the synergies.

# 10

## The Sales Force

### 10.1 SALES TACTICS OF THE MASTERS

A harvest is the product and reward of effort. The harvest of engineering, production, and sales activities effort is gathered in the marketplace—provided that the sales force is motivated, well-trained, believes in what it is doing, and knows how to serve the client.

Sales is a creative activity, though at times it goes into excesses. Sales training must be polyvalent, but also well focused. The human contact is paramount. Innovation, quality, and price, however, also play a very important role in the constellation of reasons that make the customer kick.

Globalization intensified the importance of pricing. The native going to the market to sell his goods has no set price parameters because his product costs him, primarily, his personal labor. Modern industry cannot afford such simplicity. Both materials and labor are costly factors. Reductions in overall cost—particularly in materials and processing—depend on engineering:

- Cost trimming is realized through production processes.
- Effective sales require a robust distribution system to channel the goods to the market.

Market research, particularly the one conducted through direct customer interviews, is an activity in which engineering ideas find an anchorage. Studies conducted to determine customer likes and dislikes can be instrumental in informing designers where to initiate technical programs for new products and improve existing ones. Information obtained through market research also helps to stabilize product plans.

At the same time, this information serves as the basis of *sales planning* helping to classify potential customers according to types of products required, industry and location; size up the importance of *customer classes*, in terms of drives and buying power; and identify the nature of the purchasers' consumption problems and requirements. They can also be instrumental in

- predicting when and how customer needs most likely arise,
- learning the customer's methods of selecting suppliers, and
- providing the basis for effective customer handling over the longer term

The recognition of underlying factors in product sales has been considered by many as the crucial issue in market research. Both industrial firms and financial institutions realize that they must take the customer's attitudes into account, and that the better they understand the people who make up the markets the more effective their promotional and sales endeavors will be.

Done at marketing level, a rigorous sales analysis supplements the long-range aspects of projections and estimates by the sales force. In the course of our meeting, the chief executive of a major industrial concern commented as follows on this subject: "In my company, current and longer range market trends are studied constantly to help us distribute our products where and as they are needed."

This sort of input is absolutely necessary, but it is a complement and not a substitute to learning sales tactics and practices by osmosis by acting as "assistant to" and by watching great salesmen. Typically, great salesmen also have organizational and managerial qualities. Wal-Mart's Sam Walton provides a first-class example.

One of his gifts was that of anticipating where things were headed, coupled with the ability to understand the implications of the social and demographic currents that were sweeping America in the 1950s and 1960s. Together with hard work, that acumen hastened his rise from humble proprietor of a variety store in Newport, Arkansas, to largest retailer in the world.

Walton has made an important contribution to salesmanship by showing that human well-being and making money are not inconsistent or incompatible. He most likely came to this conclusion because he was a lifelong learner and paid attention to people, realizing that

- The potential of any enterprise hinged on giving subordinates the maximum opportunity to succeed.
- This is one of the basic characteristics of business success not only in sales but also in all management duties.

Walton did not invent discount retailing, just as Henry Ford did not invent the automobile. But his extraordinary pursuit of discounting revolutionized the service economy and changed the philosophy of American business. When he died in 1992, he left a broad and important legacy in American business, by the great instinct for seeing a market before it happened—and then making it happen.

At the same time, however, Walton paid a great amount of attention to the consumer and sought ways to increase demand and reduce prices. This dual accomplishment is another great sign of salesmanship. He even forced the competition to embrace this concept, making the market grow while creating a better profit stream.

When the concept of aggressive discounting began to emerge in the early 1960s, Walton was a fairly well-to-do merchant in his 40s, operating some 15 variety stores spread mostly around Arkansas, Missouri, and Oklahoma. These were traditional small-town stores with relatively high price markups, but as master salesman *and* active student of retailing he saw what was coming.

- Once committed to discounting, he began a crusade that lasted the rest of his life.
- His aim became to drive costs out of the merchandising system.
- *Cost control* was established as a strategy, squeezing margins in the stores, in the manufacturers, and in the middlemen.

Another of Sam Walton's legacies has been his belief that salesmen and firms should seek not simply to position themselves within their existing markets, but they should also capitalize on their advantages to redefine markets in their favor. Reaching that goal evidently involved

- identifying and developing strengths that cannot easily be imitated by competitors, and
- stretching the salesmen strengths to the maximum, by leading them into setting for themselves ambitious industry-transforming objectives

To cut his margins to the bone, Walton put together the strategy that Wal-Mart grow sales at a relentless pace. Realizing he cannot go far enough without high technology, in 1966, when he had 20 stores, he attended an IBM school with the aim to hire the smartest person in the class to computerize merchandise controls. With this, Wal-Mart went on to become the leader of just-in-time (JIT) inventory control and sophisticated logistics, using information technology as a competitive weapon.

But while gaining such a competitive advantage, he did not allow cost control to fade out of the radar screen. In fact, his ability to harness information technology enabled him to further cut costs. Quickly this strategy engaged the whole organization because it was a basic background factor of all management decisions. Practical examples are Wal-Mart's

- low-overhead, and
- fast flow replenishment of inventory

The Internet became the friend of this philosophy, by promoting the efficiency of business partner chains, years after Walton had passed away. Great salesmanship is in no way limited to signing up a wealthy contract. As the Wal-Mart case study demonstrates, it is a complex and polyvalent longer-term process that uses technology but also opens the door to the client's mind and heart.

## 10.2 RESULTS EXPECTED FROM THE SALES FORCE

It needs no explaining that the greater is the competition and the level of a product's sophistication, the better trained and supported should be the sales force. Globalization adds to this requirement because it has thrived on open markets and most open markets, Jim Barsdale once said, evolve or devolve into the old 4-2-1 strategy. *If* it is a big market and it's open, *then*

- the first guy gets a four
- the second gets two, and
- the third gets just one*

* Information Strategy, November 1996.

Companies that trail this list more or less come and go particularly if they do not have a really successful business model and financial staying power. This sees to it that over time, truly open competitive markets tend to slim down to two or three competitive firms that are more or less (but not always) viable.

It comes as no surprise that business landscapes that are open, competitive, and trade in hot products are, or become, big enough for people and companies who want to get into them. But at the same time they may not be big enough for those who want to stay on the market unless they are the leaders and know how to hold that position—a task involving

- engineering,
- production,
- logistics, and
- sales skills

To stay ahead of the curve, it is necessary but not sufficient that a company invests in research and development. If it does not cut costs, the way Sam Walton was doing (Section 10.1), and steadily train its sales force in ways and means needed to get results, it will lose its dominant position.

Salesmen must be carefully selected and given a good schooling in order to rely on them to promote products in the market. Training must help them answer challenges with which they are presented to push the products they sell. And they should be convinced that there is no better alternative than dancing with the company that brought them to the ball. A sound training program in computers and communications will typically involve

- the fundamental notions of good salesmanship
- complete and thorough sales-oriented information on the product(s)
- how to offer these products successfully, and how to sign a sales contract
- basic skills to become an able "discussion partner" with the client's experts or purchasing agents
- case studies to prove how a first-class sales job is being done, and how to help the customer in implementation of the wares
- customer meetings and training sessions, including demonstrations of successful applications, and
- a good library of references and a variety of assistance aids (like XSELL and XCON, Section 10.4)

Training must not only improve the salesman's technical know-how, but also his or her ability to channel the product(s) to the market. Moreover, this ability should be upkept through information meetings and a steady stream of literature on competitive intelligence. The latter issue—which has been for decades a basic IBM policy—is not always appreciated by management yet it is an important contributor to above average results.

Each salesman's, sales office's, and territory's results must be constantly evaluated against targets (more on this in Section 10.3). Left on their own devices the sales force's deliverables dwindle. The effort loses its steam. Table 10.1 provides real-life statistics from the auditing I did several years ago in the branch offices of a data systems company. The branch employed 10 salesmen whose productivity varied widely from plain loafing to acceptable results.

The Vice President of Marketing of the data systems firm in reference expressed himself in this manner. "With the lack of the right products and without the adequate salesmen, this company's share will never account for the full market potential. We must realistically consider how many data technology products a salesman can sell."

**TABLE 10.1**

Sales Statistics on Data Systems Products in One of a Data Systems Company's Sales Offices

| Salesman | Yearly Sales ($) | Quantity Sold in Units | Average Price ($) |
|---|---|---|---|
| A | 1,800,000 | 38 | 47,368 |
| B | 504,000 | 63 | 8,000 |
| C | 248,000 | 12 | 20,666 |
| D | 847,000 | 93 | 9,107 |
| E | 24,000 | 8 | 3,000 |
| F | 300,080 | 27 | 11,111 |
| G | 97,800 | 6 | 16,166 |
| H | 255,000 | 20 | 12,750 |
| I | 687,000 | 19 | 13,578 |
| | **5,020,000** | **321** | **136,258** |

*Notes:* Average price per unit $ 15,638
Average Units per Salesman 32
Average $ per Salesman/per year 502,000
Range/price per unit 3,000 to 47,368
Range/units per salesman 6 to 93
Range/$ per salesman, per year 24,000 to 1,800,000

It did not take any extraordinary effort to find out that this firm's sales force was at a state that did not permit a high level of sales. Rather than increasing the personnel in order to sell the number of units, the company had to weed out the unable and unwilling, hire young salesmen, and intensively train the whole lot.

Critical questions, too, had to be asked. What is a salesman doing wrong if his sales are below average? Leaving aside the case of plain loafing, probably he or she was planning poorly or not at all his customer visits. Or, he lacked the proper know-how on how to manage time and territory (which is, generally, the No. 1 problem for salespeople).

Unquestionably, the onus was with the company because it failed to help its salesmen in doing a proper job—not only product-wise but also in supporting them with the best sales technology available. This happens surprisingly often. The answers to 10 questions, to which every sales organization should respond affirmatively, help in improving sales results:

1. Are sales objectives and profit objectives set for each account?
2. Do we have a system of classifying accounts according to their potential?
3. Have we made an organized study of our salesmen's use of time?
4. Do our salesmen use "call schedules"? Are they effective?
5. Do our salesmen use prescribed routing patterns in covering the territory?
6. Do we determine in advance how many calls to make on an account in a given time?
7. Do we have the information to decide how long each call should take?
8. Do we have a call report system? Do our salesmen report on every call made?
9. Do our salesmen use elements of a planned sales presentation?
10. Are we regularly revising the sales quotas (Section 10.3) as new products are introduced, and the pricing of older products change?

In addition, sales meetings by branch and by specialty (product, product line, new improved software release) help in improving salesmen productivity. This is part of a policy stating that the introduction of new products should bring together the salesmen with the product manager and product planner.

A similar statement is valid in what concerns handholding with the marketing staff. Account managers should be reporting in no ambiguous terms on what they see as competitive advantages/disadvantages of the products they sell, as well as their own ideas for improving business opportunities by

- exploiting the company's strengths, and
- swamping its perceived weaknesses

Salesmen cannot perform without full knowledge of the company's products that they handle, of their customers' drives and of the market's trends. Neither is it that existing demand for a certain product necessarily means that *we* produce is going to sell. Conversely, little or no demand for a certain product at a certain time, and in a certain place, should not be interpreted as a strong sign that it will not eventually sell. Demand can be created if we know the facts, appreciate the needs, and know how to go after the business. This of course has prerequisites:

- Business opportunity studies assisted by market research
- Business intelligence studies to analyze the actions of competitors
- Establishment and observance of quotas (sales objectives, Section 10.3)
- Evaluation of actual versus planned quantitative objectives
- New account analyses and guidelines based on such analyses
- Customer complaint investigations and quick response to them
- Return-on-sales analyses to steadily improve not only deliverables but also the cost/effectiveness of sales effort

Among important marketing and sales decisions are those concerning the reevaluation of the distribution network—its growth, operation and expenses—and the evaluation of opportunities (as well as constraints) presented by alternative sales channels. Considerable thought should be given to the validity of a dealer organization and dealer policies (should such an organization exist) against the alternative of developing fully company-controlled sales channels.

## 10.3 ESTABLISHING QUANTITATIVE OBJECTIVES

Not all companies use the same ways and means in planning and controlling their sales force. In my experience, I have found that the most frequent feedback means (by four companies out of five) is the institution of a call reporting system. Another common policy is to try to get the maximum possible business out of existing accounts.

At IBM's high times in the 1960s and 1970s, orders from existing clients represented 80 percent of annual intake. This is a high water mark but roughly 70 percent of sales orders can well come from existing accounts classified according to potential—particularly if the sales effort is well planned and managed. By contrast, if the company lacks the right sales organization sales results from existing accounts will be substandard.

As with any other management responsibility, the sales effort must be properly organized, with quantitative objectives set, and carefully watched. Quantitative measurements of sales results require the establishment of *sales quotas*. This is a "must" if *our* company is to follow a sound approach to

- planning and
- controlling the performance of its salesmen

Work on quantitative sales objectives should be conditioned by the belief that the quota system is a fundamental tool of sales management. This is particularly important in a multinational company where marketing headquarters allocate quantitative objectives, each nation's manager sets the sales objectives by sector, and each sector manager establishes quotas by salesman—to be approved through a feedback loop. All prevailing conditions such as

- type of product(s)
- market potential
- profit margins
- economic indicators, and
- availability of a trained sales force, have to be taken into account

Thomas Watson, Sr. was a master of the quota system. Accounting for different marketing conditions by country, IBM had divided the individual sales territories around the world into "fertile" and "poor." At the time of this reference a fertile territory runs at 20,000 points per month rental per salesman (a point equaled one dollar of monthly rent for the installed equipment). A poor territory stood at about one-third of this amount.

The territorial distribution was so arranged as to give overall between 15,000 and 18,000 points per month per salesman in the average, or better than 200,000 points per year. This has been a balancing act raising the question, "What guidelines should be followed to keep it going?" In the course of a project aimed to restructure the sales effort, at a well known computer company I was consultant to the general management, we established that five basic considerations should be considered for the establishment of individual sales quotas:

1. Product mix of the equipment salesmen are required to channel to the market
2. Orientation of the sales effort, particularly in what concerns industry marketing
3. Penetration of the company into a certain market and the fertility of that market
4. Quality history of a salesman, including his "business age"* and personal characteristics
5. The nature and extent of sales support (Section 10.4)

In establishing these criteria, we accounted for the fact that, as every salesman worth his salt has found by experience, a sale depends on actual demand in the market, on the product itself, on the sales promotion effort, on competition, and on the ability of the company's sales force to bring its message "close to the heart" of the purchasing manager or consumer.

To help in establishing quantitative objectives, we took a leaf out of Big Blue's book. As mentioned in the preceding paragraphs, originally IBM calculated a salesman's commission on the basis of the point system. For new accounts, the commission was higher. When rented equipment was being taken back, its rental value was flatly subtracted from the rental value of the newly installed equipment and commissions were only paid on the difference.

Computers are no more rented these days, at least not in any significant way, and there are as well other changes required to adapt that system. Quotas and commissions have become more complex as the products have multiplied, while the unit price (of measures such as byte of storage or instructions per second) has dropped dramatically. The market has changed, but issues like salesmen productivity continue to be a crucial factor.

* Number of years on the job.

The computer manufacturer mentioned in the previous paragraphs chose a basic salary-and-commissions approach for salesman compensation. The "typical" sales engineer had half of his salary fixed and the other half on commission based on his quota—and he received half the commission on signing the contract and the other half on installation.

A really good sales engineer, however, could gain through commissions three to four times his basic salary. This was deliberately adopted to promote effectiveness in sales. Pushing the product to the market is a particular concern to any manufacturer of high-technology equipment.

Precisely for this reason, the aforementioned system of quantitative sales objectives was structured to tie the salesman's commission to deliverables. In addition the system provided for continuity in the employment of the company's salesmen. With this goal in mind, another company I was associated with adopted the following approach: A quarter of the commissions were credited on signing the contract, another quarter upon installation, still a quarter at the end of the first year, and the last quarter at the end of the second year.

- The latter two percentages were known as "maintenance commission."
- *If* a salesman left prior to receiving the maintenance commission, *then* his share was distributed among the remaining salesmen.*

To improve performance, the company's salesmen were provided with a growing armory of technological supports, some of which targeted sales and others helped in customer handholding. The policy of sales support must always be forward looking, especially by a company marketing technology-based systems.

Further support was provided through systems skills located at the branch offices, to help promote the meeting implementation of sales quotas and answer specific products queries posed by customers in need of technical assistance. In exchange, the company asked its salesmen to base their effort on convincing the clients about the merits of the wares being marketed—rather than on hard sales.

This decision was taken in reflection of the fact that regardless of the informal, friendly attitude conveyed by the vendor's salesmen computer marketing

* This proved to be a good incentive. Salesmen were not adverse to it because commissions were higher than those adopted by competitors. The original idea of a time distribution is credited to Jim Allen, founder of Booz Allen Hamilton.

efforts are neither informal nor friendly. They are tactical exercises involving not just detail and sales force discipline but also arm-twisting.

The sales program pressed the point that the fact the user organization generally trusts the vendor and its representatives, should not be employed to take advantage of it. Instead, emphasis was placed toward building rapport and establishing a relationship with the firm, its CIO, and his end-users.

## 10.4 TO BE AHEAD OF THE CURVE, USE KNOWLEDGE ENGINEERING, NOT ARM-TWISTING

One of the leaders of industry who used technology to improve sales and income has been Ken Olsen. In the mid-1980s DEC was the first computer manufacturer to develop and use an expert system called XSEL (Expert Salesman). At a University of Vermont conference at which we were lecturing together, I asked DEC's EVP of Marketing what motivated that development and he answered, "At DEC we hire the best engineering graduates and put them in sales. They know everything on VAX design, but not in salesmanship. XSEL fills the gap, assisting them in sales."

Also in the mid-1980s DEC had developed its high water mark expert system. Known as Expert Configurator (XCON; this was a terrible and complex) knowledge construct. Originally written by Carnegie Mellon University (CMU) as Project R1, it reached 500 rules, grew to 800 and was then transferred to DEC which added to its sophistication. Including associated subsystems, when fully developed XCON

- featured over 8000 rules
- represented a significant investment, and
- managed more than 12,000 VAX/PDP components

*If* these programs were in Cobol, the obsolete legacy software programming language (that still ravages those companies who unwisely continue to use it) *then*

- it would have been able to do only 1/3 of XCON's job,
- would have involved an estimated 1.2 million to 1.5 million Cobol statements, and

- would have taken 200 to 300 man-years of effort, at over three times the cost of the expert system solution

DEC's leadership in advanced technology at the time XSELL and XCON were written, makes interesting reading because it shows how a great engineering company can transfer skills from design to sales. There is a strict engineering methodology implied through the development and use of XCON—primarily a sales support tool (while XSELL assisted in sales proper. At a meeting, DEC gave six reasons why it developed and uses expert systems. They

- make feasible high service quality,
- assure knowledge distribution,
- contribute to greater productivity,
- are entries into a growth market,
- competitors are also coming into this field, and
- they are one of the best means for swamping costs

In terms of cost-avoidance after it went into service XCON represented for DEC $30 to $40 million per year in savings from not having to do VAX configurations the old way, then discount them because of different errors or omissions—losing time and money in the process. After the knowledge artifact established a name for itself in marketing and sales, other companies too started to develop XCON-type system configurators. An example is Reuters. Its expert system's mission was movement and delivery of financial information terminals and other equipment around the globe.

One of the benefits of being ahead of the curve in sales technology and methods is that using companies gain confidence in themselves, serve better their clients and don't have to resort to arcane arm-twisting in order to sell their products. They train their salesmen on how to overtake their competitors and how to handhold—rather than on how to corner their clients, by making false promises or by paying bribes.

This practice of *explaining* and *leading* is not necessarily widely adopted with the sale of computers, electric engineering gear, aircraft, and other man-made systems—let alone being a generally applied standard. Instead some companies in the computer industry capitalize on the fact that senior executives of user organizations do not have adequate knowledge of hardware and software, or the time that needs to be devoted to contract negotiations.

- They are burdened with a variety of other pressing business matters.
- They seldom ask the vendor for verification or written confirmation of sales promises and of negotiated details that should be incorporated to protect the user's interests.

In the case of the computer vendor whose case study was the Section 10.3 theme, the board of the company decided that when confronted with disloyal competition, its salesmen would not join in, but they would bring it to the attention of the user organization's responsible executives. They would also explain that such ploys come in all shapes and sizes and in varying degrees of abusing the client's confidence.

For instance, one of the ploys in not-so-ethical sales of computers involves excessive hot air and misrepresentation by omission of important facts by the salesman. Another example is the salesman's refusal to incorporate in the written contract the promises he has been making to his client.

In the case of an Italian bank for which I was consultant to the president, we asked the institution's traditional mainframe supplier (name deliberately withheld) to cancel the ongoing contract and sign a new one prepared by *our* legal advisor—in *our* terms. This new contract prescribed specific penalties if the computer maker repeated his past arm-twisting policies, or invented new ones.

In what became known as the case of "a basket of tomatoes," one night a truck belonging to the computer vendor drove up at midnight at the bank's computing center. The man in charge of the night shift was a junior-manager and had no signature rights. Without any authorization on the bank's behalf the vendor's driver unloaded the crates of a big mainframe and asked the night shift supervisor to sign a receipt.

The junior manager first refused, but the driver insisted, telling him, "Just sign as if I delivered a basket of tomatoes." Under pressure, and incorrectly, the supervisor signed the receipt. The next day we called the mainframer to come and get his wares we had not ordered. Not only did he refuse, but he also sent a rental bill for the "basket of tomatoes."

This brought our lawyers into action. By a registered letter asking for receipt, we informed the mainframer that the earlier contract (held in the vault of the president) authorized him to publish and publicize the twists, in the event that something inordinate happened to the vendor. The mainframer then sent his lawyers to attend a senior management meeting. We welcomed them. Having done our homework by way of legal discovery, we were ready for them.

The mainframer's lawyers were dismayed. They simply did not expect what they saw and read at that meeting. Usually computer user organizations just bend over when the vendor gets tough. Afterwards the mainframer not only collected his "basket of tomatoes," but also agreed to pay a hefty penalty. Following that we became "amici come prima."

The morale of the story is that the user should never, ever sign the vendor's standard form of a contract. He should always involve his attorneys early enough in the negotiations, to allow them to prepare a meaningful contract that protects the user organization's interest in the transaction. And, most importantly, the user organization should have the guts to challenge the computer vendor when he misbehaves.

## 10.5 BRAND RECOGNITION

Lipstick, eyeshadow, and makeup are terms that were popularized by Max Factor. Their power over consumers lies in the fact that they are composed of simple words, easily understood, and remembered. Since the start, branding has been based on ingenuity and catch words easy to remember.

"Makeup in seconds, look lovely for hours," was the headline in a typical Max Factor advertisement. For millions of ordinary women *Factor* simplified the complicated task of looking, or at least feeling, good. In the process, it also helped to create the first major international cosmetic business.

Nowadays in merchandizing brand recognition and brand management is seen as the way to generate the nearest thing to strong identities—whether the product is cosmetics, cars, cornflakes, or toothpaste. For companies in diverse industries like microprocessor chips, computer products, minivans, or packaged food brand management is cornerstone to their policy of

- stopping to chase their competitor, and
- focusing in chasing their competitors' customers

Successful branding is done in a coordinated manner. Typically, the brand manager's pay is linked to the brand's success, while advertising, pricing, and planning are each handled by a different person but in a way providing synergy.

The people in charge of a branding effort should be keen to avert the risk that competing brand heads end up brawling among themselves as they vie for top management attention and marketing money. This leads to cornering the marketing effort, and such corners led P&G itself to abandon brand management in 1988.* But other companies persisted, and brand recognition spread to many industry sectors.

Brand recognition is just as important in high technology. Intel, Cisco, Amazon.com, Apple, Microsoft, and IBM share the distinction of being ranked among the top Fortune 500 companies by their globally recognized *brand*. "Intel Inside" has been one of the most successful slogans. Financial analysts say that after Intel's success technology companies are obsessed with getting brand recognition.

Yet, when in 1992 Intel began its branding campaign it was, at first, met with skepticism both outside the company and within it. Intel's customers were PC manufacturers who did not need a fancy brand to tell them what kind of performance and value they were getting from the company's chips.

That did not deter Intel's management, which also had the bright idea of contributing directly to the PC makers' marketing campaigns, as long as they promoted "Intel Inside." Before too long, the chip maker's footprint could be seen all over the market, outstripping that of any PC manufacturer.

- "Intel inside" became a *trustmark*.
- The trademark was raised to a level that consumers regularly put their faith in.

Not only in the United States but internationally, technology firms feel that serious brand-building is important because it projects them in the public mind, all the way to the top of user organizations. For example, major information technology contracts are awarded at board level, rather than by IT managers alone, and directors tend to question the wisdom of signing up with little known companies.

---

* Deciding that its system was pitting brands like Camay and Ivory soaps against one another, P&G switched to "category management" in which managers oversaw whole product categories.

Even well-established technology firms express the opinion that their future will increasingly be influenced by boards that operate far beyond confines of the technology industry. Brand recognition can help to win the support of nontechnical executives, who know little about IT, raising the barriers to smaller, less known rivals—as IBM once did.

Branding can also serve as a Trojan horse. As the twentieth century came to a close (specifically in 1997), AT&T started to consider a novel and less costly way of entering the telecommunications market—doing so more quickly than building its own infrastructure. The strategy was to franchise its brand name to local and wireless carriers, directly connecting to the AT&T network and billing system.

The plan was that branding would enable AT&T to quickly bring wireless and local services to long distance clients, as well as save it up to $3 billion from its $9 billion capital budget. This branding strategy contrasted, for instance, to MCI which at the time operated cellular service in 50 major markets although it did not own the facilities. MCI's strategy, which also used brand name, was to

- buy transmission from local companies, marketing it as an MCI service, and
- bundle it with other business services to create an appealing package

In the background of these strategies was the fact that old telecommunications standards policies were turning on their head because of wireless services spreading all over the United States and rapidly expanding the potential market area. Through branding, Telcos hoped to expand rapidly with little capital outlay, but their management was wanting and franchising plans did not provide expected results.

Financial analysts said at the time that at the end this branding experience damaged the firm's reputation. AT&T, for instance, was a widely respected name for high quality services and exceptional customer care—but as its franchises failed to uphold the high standards expected by customers, AT&T was hit. The lesson is that branding and franchising management needs to

- exercise steady vigilance, and
- pay a great amount of attention to quality control

Cornerstone to their marketing is the development of a brand enriched by a message, "You have a friend." (Chase Manhattan was the first to use that concept). Therefore, they drive for own-brand products and look at them as the means to produce improved margins while increasing pressure on competitors.

Of course it is not enough to invent a brand, or a slogan in order to experience zooming sales. Moreover, the sort of brand that hits the public eye does not come by revelation. It needs careful study and testing, including market research and contrarian opinions. This does not happen as a matter of course.

In a number of cases, all sorts of industries searching for branding industries and banks are actually in the dark as to how their products or services are regarded by present and potential customers. Management is not always aware of this, yet it has to make decisions on product design, packaging, pricing, and promotion which

- entail large sums of money, and
- in some cases determine the success or failure of the enterprise

It is true that a certain amount of hit-or-miss, trial-and-error tactics is inevitable. Consumers do not always know why they do what they do; they tend to be all different from one another, and even the same people vary in their feelings from time to time. The readers of a serious business magazine may think they subscribe to it because it is thorough or authoritative, whereas they really take it because it makes them feel important.

Some people will say they like a brand of beer because it is "dry," although they actually prefer it because by habit they have been accustomed to it. Others might say they dislike a certain brand of cigarettes because it is "too strong," but the underlying reason may be that they associate it with heavy smokers.

In making decisions on the basis of factors that are inherently slippery and intangible, there is a clear need for information that will add some degree of definiteness. In addition, any company in which customer motivation is crucial will find that being even a little closer to the target can produce clear-cut advantages in the sales picture. This leads to a variety of approaches in making brand-related estimates.

In conclusion, the significance in branding lies in the nature of the effort being made: To penetrate to a deeper, less conscious layer of consumer or company thinking, where the real motivations for buying this or that product may lie unrecognized. This clearly contrasts the logical-sounding reasons that consumers often believe (and say) they have, and that can lead a marketer to the wrong moves if he accepts them at face value.

## 10.6 SALESMANSHIP AND ENTREPRENEURSHIP CORRELATE: REICHMANN AND THE CANARY WHARF

The year 1992 saw the Canary Wharf collapse. Practically everyone said that in deciding to build a new financial center in London's Docklands, the Reichmanns underestimated several factors: the logistical problems in getting there, the self-preservation instincts of the City, and the conservatism of the British business classes.

In other terms, this was a failure of salesmanship rather than of entrepreneurship of which the Reichmanns had given plenty of evidence in North America. As one commentary had it, the Canadians made the mistake of equating London with Toronto and with New York. In The Big Apple, their World Financial Center was a stunning success, but

- the World Financial Center was only five minutes' walk from Wall Street, and
- the Canary Wharf is a half-hour taxi ride from Threadneedle Street and much longer in the rush hour

Another argument has been that Wall Street did not, and could not, react to the Reichmanns' scheme by building another dozen skyscrapers of its own. London's City did exactly that, constructing giant new developments such as Broadgate, Cannon Bridge, and Alban Gate, resulting in a glut of office space for financial companies.

Financial analysts and commentators asked, "Why should anybody move to the Docklands when there are dozens of vacant developments at bargain prices in the center of the capital?" The conclusion was that they shouldn't, and that the banks must bear the blame for miscalculating the response of the British financial and business establishment.

The pros answered that all that was garbage. The critics simply forgot that Reagonomics had led to the go-go mid- to late-1980s on both sides of the North Atlantic. In the United Kingdom, Barclays became the most aggressive lender during that period, and the period from 1987 to 1991 saw its loan portfolio to the real estate sector grow by 35 percent a year. It was the banks who misled the Reichmanns.

The banks compounded their errors when they came to Olympia & York Development (O&YD) lending hundreds of millions of dollars on the basis of little more than the Reichmanns' good reputation. In the case of Canary Wharf, the credit institutions did not even demand cross-collateralization on O&YD's valuable North American properties. As a result, their only security on these loans, to the tune of £1.2 billion ($1.75 billion, at the time) was practically worthless, and huge losses were on the cards.

According to this argument it was the banks' entrepreneurship and not the Reichmanns' salesmanship that was at fault. Indeed, many of the largest banks in North America, Europe, and Asia lent lavishly to Olympia & York Developments—which was at the time the world's largest private real estate company. Many analysts said they did so by violating their own credit rules because O&YD

- would not let the banks review its books, and
- refused to disclose all its financial obligations

As a result of this glut of money thrown at it, nobody was surprised that O&YD was snowed under too much debt while facing depressed property markets. This amounted to a series of disastrous investments, and the banks were nervously lining up to see what their losses were likely to be.

Those who thought that the Reichmanns' salesmanship was still stellar, pressed the point that the banks bet on the fame of the Canadian family, rather than on reason based on analysis. This was not so unusual because after decades of closely guarding its finances, Olympia & York had always the upper hand. Only after the 1992 disaster it was forced to agree to open its books and show its creditor banks

- who owed what, and
- on what sort of terms

One of the basic principles of credit is that unsecured lending without knowing the full financial picture is very dangerous. By contrast, secured lending based on the value of specific projects is reasonable even without knowing the full financial picture. Every banker knows that principle, and nearly everyone violates it.

The credit problem at Olympia & York is that it was carrying $15 billion in real estate debt and more than $3 billion on its oil and gas subsidiaries (a big amount of money, at the time). This the banks learned only after adversity hit. Like Greece, Spain, Portugal, and Ireland nowadays, the company was facing a string of large debt payments that it could not meet. Therefore, it was pressing its banks for

- even more money in short-term loans, and
- much more lenient terms on its existing debt

Of all the bankers' reputations to have been hit by the Canary Wharf debacle, none was damaged harder than that of Sir John Quinton, chairman of Barclays Bank. Till then, Quinton was lauded as the man prepared to change the tight lending policies set under his predecessor, Sir Timothy Bevan, in order to reestablish Barclays' leading role in British banking. But as the dust settled around the fallen Reichmann empire, Barclays and Quinton appeared in a totally different light.

Several analysts also suggested that O&YD had not been Barclays biggest credit pain. The bank's single largest property exposure was £440 million to Imry Group and its parent, Marketchief. Imry and Marketchief were forced to refinance their bank debts in 1991, but signs were that they were pulling out of trouble. Between 1987 and 1991, Barclays' total exposure to the property sector

- had increased from £2.1 billion ($3 billion) to £5.4 billion ($7.73 billion) a 157 percent increase, and
- this compared badly to a 64 percent rise in total United Kingdom lending for the same period

Olympia & York owed Barclays £200 million (then $290 million) including loans to Canary Wharf. There seemed little prospect of Barclays getting much of this back, and even within the bank questions were being asked about how much a supposedly conservative institution should have lent to so much of a secretive company. This led to a top management shake-out.

In the aftermath, John Quinton announced he would step down as chairman at the end of 1992, to be replaced by Andrew Buxron. Despite the bank's denials there was as well persistent speculation he was forced to go amid boardroom dissatisfaction at the bank's growing problems. The management of risk is an art that very few bankers, including CEOs, master.

## 10.7 DEEPER AND DEEPER IN DEBT IS POOR FINANCIAL SALESMANSHIP

Acting as intermediaries, banks make a big chunk of their profits from loans, after all this is their prime activity as financial go-between. But pushing loans in a blitz, which means without proper consideration of credit credentials, is very poor salesmanship and a prescription for hard times. Sooner rather than later it lands into trouble

- the bank's clients, and
- the credit institution itself

As loan amounts spiral while creditworthiness shrinks the board may take steps to correct this situation. Usually, heads have to roll. But the damage may have already been done, and quite often those responsible are no more around.

For instance, when the Reichmanns' financial troubles broke out, several of the directors responsible for Barclays Bank's rapid expansion in the property sector were no longer at the institution. Of the eight key executives pictured in the 1987 annual report only five remained; so the chairman paid the price.*

Among American banks Citicorp was believed to have lent about half a billion dollars to Olympia & York, Reichmann's vehicle, but rumors were that it had sold pieces of those loans to other banks. Chemical Banking was believed to hold $250 million of loans to O&YD. The irony was that some of Olympia & York's creditors were the same banks that lent Donald J. Trump $2 billion without reviewing his financial records, or his property appraisals. Those banks were stunned in 1990 to find that Trump

- had overborrowed,
- was short of cash, and
- his property values had fallen most drastically

One of the credit policy facts to which bankers lending money to O&YD did not pay attention was that the signs of trouble at the company have

* Barclays was able to weather the storm, while other credit institutions overexposed to real estate went under—like the Bank of New England, in the United States.

been emerging for at least two years. While the depth of the abyss could not be measured, because the company's finances were so closely guarded, it was not a secret that the clock was no longer functioning as before.

This carelessness in extending credit is by no means an exception. Banks specialize in competing for loans, and to do so they simply forget about the client's or prospect's creditworthiness. In O&YD's case executives of wounded banks were quick to say that the company always vehemently denied it was having difficulty, even when in 1990

- it sought to raise cash by trying to sell 20 percent of its U.S. property portfolio, and
- when that did not go through it tried to sell its oil and forestry businesses

The fact that both sale attempts were unsuccessful should have rung alarm bells. It did not. Then, in June 1991 O&YD borrowed $160 million from a group of banks, sold property in Florida, sold assets of its pipeline business, and got the Hong Kong billionaire, Li Kashing, to buy a stake in one of its troubled buildings.

To the extent that borrowers approached the banks and the banks did not exercise due diligence in credit risk control, is evidently the banks' fault—the wrong sort of salesmanship. The common problem of real estate developers is that they manage to borrow heavily against properties at the height of the real estate market, which is their kind of exercising salesmanship. Then

- the values of their holdings go down sharply, and
- the properties are worth much less than the billions of dollars in debt against them

According to anecdotal evidence, in the O&YD case many of New York's top real estate analysts had executives privately say that there was little or no remaining equity value in the company's New York portfolio. This was a direct result of the fact that Olympia & York financed its $3 billion equity contribution to Canary Wharf largely by remortgaging some of its New York buildings.

"The thing about Canary Wharf," said a credit risk expert in a meeting, "is the more you look at the numbers, the worse the whole thing looks. You can turn the numbers upside-down, sideways, horizontal and vertical and it just doesn't look any better. There is no way, no matter how you twist it and turn it, to change that."

Even the mathematics seemed to be wanting. An elementary calculation showed that if the nearly 5 million square feet built were fully let at £15 a square foot, a rather optimistic assumption in that kind of market, Canary Wharf would generate £75 million a year. The trend in London City rents and Canary Wharf rents was just as bleak, as shown in Figure 10.1.

Compared to that, the banks' exposure was £1.2 billion, pushing the interest bill to at least £120 million. In other terms, even if it were full Canary Wharf would face an annual deficit of at least £45 million or more. This would have still left thirsty the lenders who were split over whether to pull the plug on Canary Wharf.

- Four of them, Citibank, Crédit Lyonnais, Crédit Suisse, and a Finnish bank, wanted to extend more money.
- Seven, including Lloyds, Barclays, and the big Canadian banks had said "no deal."

As subsequent events were to show, the "nos" were wrong—but at the time the prospects for survival were looking slender, while lawyers were poring over leases to see if tenants, such as American Express, could extricate themselves from their commitments to Canary Wharf thereby making a bad situation even worse.

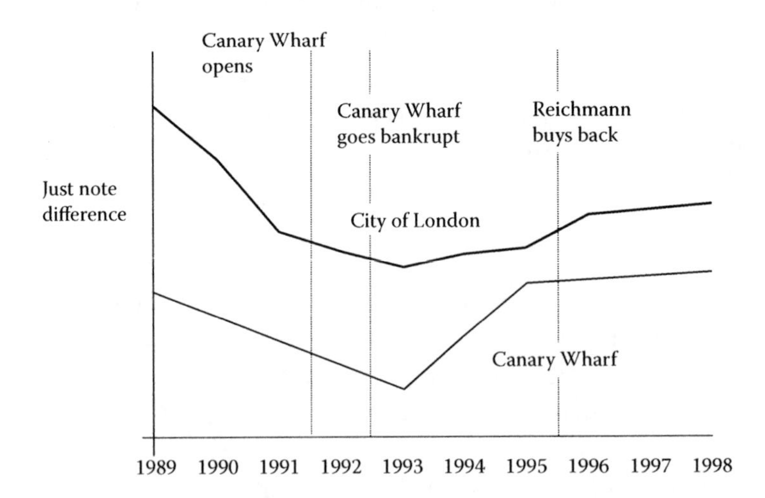


**FIGURE 10.1**
Trend lines in average rents, in £ per square foot in the City and Canary Wharf.

For its part, the British government had not decided whether to move thousands of civil servants into the project, and hopes for the £1.4 billion extension to the Jubilee Line, the underground that would link the development to central London, were yet far from being met. (Eventually, the British taxpayers put up the money for the extension.)

There was however a best-case scenario involving a big new investor. Hope was that Hong Kong billionaire Li Kashing would buy into the project, pay off the £1.2 billion to the creditors and complete the scheme alongside the O&YD management team. By contrast, the worst-case scenario had been that no backers can be found, Canary Wharf is mothballed and its elegant boulevards are left as a corporate ghost town. Reality came somewhere in between, and the man who finally saved the project, and benefited from it was none other than Paul Reichmann.

As an *Economist* article had it some five years after the downfall of Canary Wharf, Paul Reichmann was adamant that his Canary Wharf enterprise was not the source of his downfall. For this, he blamed his expansion into nonproperty businesses.* He was so sure that the Canary Wharf project was worth its salt that in 1995 he led a group of investors that bought Canary Wharf back from the banks for £800 million (then $1.12 billion). This was a paltry price compared to investments, yet above the next bidder. By 1995, the prospects were again looking good, but the banks were reportedly still feeling like being in wet pants. The existing 4.5 million square feet development was 80 percent full, and Reichmann was beginning the next phase, which could add as much as 10 million square feet with a new building already pre-let to Citibank.

A year later, in 1996, came the extension of the underground railway, and Canary Wharf became only 15 minutes from London's West End. Above all, with 19,000 people already working in Canary Wharf, the project had gained momentum, while the offices it offered were far more spacious and modern than anything in London's City.

Paul Reichmann himself had changed. He was no more the superactive entrepreneur. He had only a five percent stake in the new venture, and had embraced a more boring form of property development, in which buildings were built only if they were substantially pre-let. But Reichmann had managed to transform the skyline of London, his towers forcing other office owners, particularly in the City, to upgrade their premises.

* *The Economist*, May 25, 1997.

Postmortem some analysts said that the 1991–1992 recession, which hit Canary Wharf and the Reichmanns, was compounded by misplaced ideological fervor by the British government. One fundamental lesson of the Canary Wharf disaster was that major urban developments cannot be left to the private sector alone. Individual firms, however large, have neither the incentive nor the capacity to provide railways, roads, and other collective services.

It is a misinterpretation of sensible free-market economics to suggest government has no role in economic development. Rather, want it or not the state is on the book to provide the basic infrastructure, especially transport links.

The La Défense development in Paris, the French equivalent of Canary Wharf, can be taken as an example of this principle at work. The French government spent heavily on underground and rail links, and sent thousands of civil servants to work in the new office developments. As a result La Défense, which hosted the 1989 Group of Seven economic summit, became the modern office space breakout of the French capital—and it offered a golden horde of office space for sale to real estate agents.

# Part Four

# Innovation

# 11

# *Technology*

## 11.1 RESEARCH AND DEVELOPMENT

In its broadest sense, *technology* is the flow of new ideas that are being researched and, if promising, enter a stage of development aimed to turn them into novel and useful products. This flow of ideas, and of technology's output, is the only way of getting annual growth rates that sustain and improve the peoples' standard of living.

From the beginning of the recorded history, the progress of mankind has been characterized by certain outstanding advances in science and technology. Discoveries and inventions were always the oil greasing the wheels of progress. In modern times, which practically means since the mid-nineteenth century, advances in technology have become the scope and effort of research laboratories.

- Organized R&D often takes the headlines.
- But at the same time, private research initiatives have been major contributors and may continue doing so.

*Organized research* can be defined as a planned program directed toward the search into what is still the "unknown" in physics, chemistry, biology, and other sciences. The aim is the test of hypotheses and discovery of new facts, new applications of accepted facts, and novel interpretations of available information.

*Development* usually refers to the application of knowledge gained by research. Broadly speaking, research and development (R&D) are processes of inquiry, examination, and experimentation for the purpose of discovering and interpreting facts, revising accepted conclusions, or investigating and challenging dominant information. This is done in order to enlarge our insight of physical processes, as well as discover new resources, products, systems, methods, and arts.

"We can do nothing without preconceived ideas," said Louis Pasteur, the great nineteenth-century French scientist, "We must simply have the wisdom not to believe in their deductions without confirmation by experiment. When subjected to the rigorous control of experimentation, preconceived ideas are the burning flame of the science of observations; while fixed ideas are the danger. *The greatest derailing of the human spirit is to believe in things because we want them to be like that.*"* It is in the tradition of science

- first to observe,
- then to understand, and
- finally to experiment with and utilize the forces of nature

Not only has man been doing this since the dawn of history but also without search into the unknown there is no progress. It is good news that the research effort is virtually unceasing. Research has to be made on materials, components, subsystems, large-scale ensembles, the microcosm, environmental megasystem within which we live and operate—and life itself.

No matter how one defines it, research is "a search into the unknown"—therefore into the future. The economy will shrink and industrial production will stagnate without research. But also the research effort will remain insignificant if its produce is unfit for going into production. From the economy's and the different companies' viewpoint

- research is not just "another expense," and
- it is the "cost of staying in business" just as production and sales costs arise from being in business

The achievements of research invite the proverbial long hard look at industry's contribution to the economic strength and the well-being of nations, even *if* said nations have not always been aware of this contribution. In the longer run, no industry operating in a competitive market can

* M. Pasteur. *Histoire d'Un Savant, par Un Ignorant,* J. Hetzel et Cie, Paris (No date. Probably published shortly after Pasteur's death in nineteenth century). Pasteur's dictum is the most devastating answer to the preaching by all sorts of religions: "Believe, but don't search (to find why)."

survive without investing in research, any more than it can be a going concern without producing goods or supplying services.

Doing away with research is tantamount to putting a tombstone on one's future. Yet, at the end of nineteenth century an economy-minded U.S. Congressman suggested that the government closes the Patent Office as a needless expense, because "everything worthwhile has already been invented and there is nothing left to incite the imagination of man."*

Nowadays, research policy, like science policy, is closely linked with other domains of politics. This is not very positive, because the special characteristic of research—particularly of basic research—is that its aims, subjects, and methods change continually; it is a matter for soul-searching, for uninterrupted questioning, which most frequently is anathema to politicians.

Moreover, political interference tends to alter the basic aim of research, which is to discover and enable new products. For instance, in the late 1950s and 1960s research became a status symbol for nations—and a reason for international rivalry among the leading powers. Subsequently, research budgets made the difference between the possession of high or low grade gray matter as a national resource. Dean Neil Jacoby aptly phrased this in our March 1966 discussion at the University of California, Los Angeles:

> R&D on the frontiers of knowledge have given American industry the muscle of the international sales engineering effort and the brains of the top scientists who came to the US from all technologically advanced countries abroad. The US must keep on investing the largest amounts of money in R&D just for the sake of attracting grey matter.

This may sound as a provocative statement. But it is better to identify the issues the way they are, rather than beating around the bush. "Technological imperialism" is a nasty and imprecise term used by some people to denote America's world leadership in technology. Nonetheless, this leadership has been real and growing for half a century after the end of World War II—though it is today challenged by some of the major emerging nations.

Brains not money are everything in R&D, but money also plays a key role. According to a study by the Organization for Economic Co-operation and Development (OECD), western Europe lags far behind the United

* A *patent* is in essence a government grant of monopoly rights to manufacture the invention for which the patent is sought. In the United States these rights run for 17 years, after which time it is assumed that the patent's object should move freely into the public domain.

States and Japan in research spending. This discrepancy in R&D investment means that Europe has a serious deficit in her "technological balance of payments." European countries are buying more technical know-how, licenses, and patents from America than they are selling to her.

In addition, though a "must," financial investments in R&D are only part of the story. *Time* is as much a resource as brains and money. Following long periods of indecision, European industry is running short of time and also lags in terms of ingenious selection of R&D projects.

- For any industry in any country at any time, the number of conceivable research subjects is practically unlimited.
- To the contrary, brains and funds—private and public—are finite even in the richest countries.
- The absence of a clear strategy on R&D and productization biases delivery timetables and weakens the benefits to be described from critical design reviews.*

Under these conditions, research cannot operate as an anchor of innovation and progress, giving confidence about success in the challenging global economic environment. Who finances R&D—the company or the state—should steadily challenge timetables, functionality, cost/effectiveness, and quality of deliverables. Furthermore, the national economy itself has to be restructured to eliminate bottlenecks and allow every member of the community to play an important role in a virtuous cycle of economic progress.

## 11.2 STRATEGIC AND TACTICAL PRODUCTS

Section 11.1 brought to the reader's attention that we live in an age where change, particularly technological, is the hallmark of progress. New and improved and/or much more cost/effective products are prime ingredients of a competitive economy as well as a means to conquer global markets. Scientific accomplishments have seen to it that research became a way of life. Along a growth stimulus to business, R&D is now a basis of economic survival not only of the big but also of the small and medium-sized enterprises.

---

* In case they are being held at all.

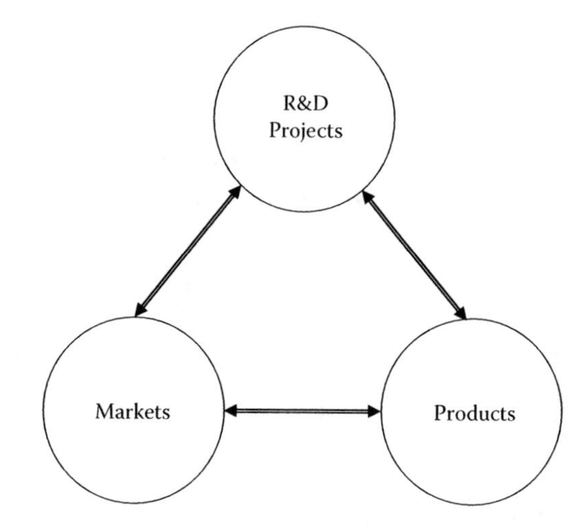


**FIGURE 11.1**
The three pillars of success in a dynamic globalized industry.

Innovation is not limited to products. Innovative ideas and processes impact any or all functions of the business, though research is especially powerful when directed to product and market leadership as Figure 11.1 suggests. Research is one of the three pillars of a dynamic, globalized industry and therefore a tremendous asset coupled to the business operations of a company:

- taking as input information on product needs,
- creating new needs through innovation, and
- providing as output new business opportunities

The message conveyed by these bullets is best appreciated in the broader concept, which states that technological developments are not just about discovery and engineering. Even the best R&D results can be sterile without commercial uptake, and deprived of market outlets they will not provide a return to required investments.

Typically, though not necessarily always, new technologies and novel products take time to attain full deployment. While we must learn to move faster selecting the right technology and then ensuring its correct application in a timely manner, processes take time and a steady effort to build up.

Product needs must be answered in an able manner, then translated into applications knowledge. Business opportunities have to be visualized in connection to new products and processes. Innovative products and services should feature lower costs* and/or better properties and capabilities

* See in Chapter 9 the case of reverse innovation.

than those already existing. A product evolution program can be designed to provide effective results *if*, and only *if*

- it has set to itself specific objectives, and
- it has been able to reach them in spite of adversities

Not all products coming out of a company's R&D laboratory are strategic. Some are *tactical* aimed to respond to short- or medium-term market requirements, or specific requests of important clients. They may also be brought to the market to provide an answer to a newly introduced competitor's product or close a gap in the company's product line.

Contrasted to tactical products, those strategic are longer term. They tend to be more generic, characterized by the fact that the company depends on them for its market impact. Table 11.1 presents the more important features that help to distinguish between strategic and tactical products.

Typically a company will have a relatively small group of carefully selected strategic products. Eli Lilly, the pharmaceutical firm is active in a broad domain of fields: neuroscience, endocrinology, oncology, and animal health; but only a few of its products are strategic. Quoting from the Eli Lilly 2009 Annual Report, "Revenue in Neuroscience, led by Zyprexa and Cymbalta, increased seven percent as compared to 2008 and represents *41 percent of our 2009 total revenue.* [emphasis added] Endocrinology, led by Humalog, Evista, and Humulin, increased three percent and represents

**TABLE 11.1**

Strategic and Tactical Products

| Strategic | Tactical |
|---|---|
| Selected few which must get most management attention; big income earners | Many products; subject to steady profit and loss evaluation |
| Longer range | Shorter range |
| Aim at market impact | Usually aimed at client response |
| More R&D, but with time-to-market deadlines | Fast to very fast transition between R&D and productization |
| Commitment to continuation | Tend to be shorter range or ephemeral |
| Commitment to steady upgrade | Frequently redesigned and/or personalized |
| No money lost in subsidies | May be given at lower cost to catch an important client or market segment |
| The degree of freedom in pricing depends on the product and its market | Must benefit from low cost production and distribution |

*26 percent of our 2009 total revenue.*[*] Oncology was our fastest-growing therapeutic area with growth of 10 percent and represents *14 percent of our 2009 total revenue.*"[†]

As this example documents, few products are considered by management as strategic, even if a pharmaceutical company launches plenty of innovative medicines that better the lives of millions of people. Those strategic help the firm to sustain growth and a strong financial performance, even if it spends a significant amount of its annual income in R&D.[‡]

The way Eli Lilly's 2009 Annual Report puts it, "The challenges facing the pharmaceutical industry today can be summed up as one fundamental problem: innovation. Our industry is suffering a dry spell in research and development at a time when society is raising the bar for pharmaceutical innovation by demanding greater value from new medicines."

Innovation in the pharmaceutical industry aims to bring to patients new, more effective medicines increasingly tailored for specific sets of problems—which means nearly customized. The way Eli Lilly puts it,

- "While this choice entails risk, it is also brimming with opportunity."
- "There's never been a more compelling case for innovative medicines."
- "As people around the world live longer, and global incomes rise, demand for innovative medicines continue to grow."

The same reference is nearly true for all products of high technology. Notice however that several problems in product evolution result from the misinterpretation of an R&D project's potential, the inadequate use of a newly developed material, sloppy timetables, and other reasons connected to mismanagement. (See also the discussion in Section 11.3.)

Plastics makers cite the polyethylene ice cube tray as a case where plastics[§] was misused. With only a twist of this flexible tray, ice cubes popped out easily. The product won wide consumer acceptance at first, but polyethylene has two basic disadvantages for this use. Since it is a good insulator,

---

[*] Idem.

[†] Idem.

[‡] To help in being competitive many companies allocate a hefty R&D budget. Osram is a lamp company. In the 1960s and 1970s I was consultant to the Osram board of management the R&D budget represented 6.5 percent of yearly income.

[§] Noun, singular.

it slows down the freezing of water; it also stiffens and often cracks at low temperatures.*

When a product flops in its intended area of implementation, a market is cut off for its maker(s). To try to limit the number of such failures, many manufacturers base their strategies on alternative product lines or search for solutions through acquisitions. Often forgotten in a rush to acquisitions is that they can lead to a clash of cultures with either of two results:

1. The labs get poorly integrated.
2. The best researchers of the acquired company leave.

Other problems, too, are present in the search for the next strategic product(s), many of them managerial. Examples include the case of top executives who want new products but do not know or cannot agree on what kinds of products to be interested in; the laboratory is crowded with development projects but with few important products coming out, and even these are not paying off; the unsettled product range, filled with potential products that have been considered for years but have never been given a marketing decision.

When these and similar events take place, the opportunity for developing strategic products diminishes to zero. A basic principle in product innovation the reader should appreciate is that no research, no technology, no new process is worth anything if it does not *create value*. Two crucial questions must be answered practically all the time in any company worth its salt:

1. How do *we* develop new technology in a way that creates value for consumers?
2. How do *we* link technology to markets so that *our* company can improve its financial staying power?

An able response to these two queries also provides food for thought in answering the next important questions, "How do we plan for and develop strategic products? How do we measure return on investment (ROI) in R&D?" As we will see in Section 11.4, R&D is a profit center—not a cost

* Most makers of ice cube trays have changed back to metal while testing other plastics to determine whether they will work any better than polyethylene.

center. And every profit center must have a fundamental framework for portraying and evaluating the return on its investments.

## 11.3 RETURN ON INVESTMENT SHOULD NOT BE TAKEN FOR GRANTED

"A man is judged by his ability and willingness to do an honest day's work," one of my professors in finance, at UCLA, taught his students back in 1953. Immediately thereafter he added, "And an investment should be judged on its return." Investments in research are no exception to this dictum.

Though searching into the unknown is an enjoyable intellectual activity, it should not be done as daydreaming. While R&D must be properly financed, it must also produce deliverables, and these should be evaluated in terms of ROI. Generalizations are always dangerous, but, other things equal, strategic products have greater- and longer-range return than those tactical. As far as creative work is concerned

- ROI is not just a matter of authorizing financing, and
- it is as well, if not predominantly so, a question of keeping the human resources on their toes, through tangible contributions to society's wealth

Whether corporate or national, many R&D plans and projects to date underestimate both the extent of available human resources and the level of personal satisfaction through intellectual effort. In addition, the number of researchers and of technically trained auxiliary personnel is as limiting a factor as money or time—and therefore human resources should be managed with ROI in mind.

The bad news is that, given the swarm of applicants, universities are nowadays more interested in numbers than in quality of output. Therefore, they are no more producing an adequate number of people with *research spirit*. The way an old Greek saying has it, "the dress does not make the priest. Neither is the label 'researcher' making a person able to investigate into the unknown." Reverting this trend will be a challenge, and it can only succeed through steady political support.

The culture characterizing R&D management and its decisions, in the majority of companies, must also change. Long decades of experience have shown that even a very intensive industrial research effort may be of little or no value in the evolution of the enterprise *if* the management fails to consider the intellectual effort behind research as

- an integrated, multiaspect, dynamic process, and
- a major contributor to the company's growth and survival

According to the National Industrial Conference Board, R&D work now occupies about two percent of the American working population. This seems small fry, but this is the wrong impression if one accounts for the fact that in R&D presently works one out of every three U.S. scientists and engineers. And although an estimated three-quarters of such research is done by private industry, roughly the same fraction has been financed with federal funds.

This is all for the better because an economy that can show both industrial initiative and a steady spirit of innovation will be dynamic and progressive. Such an economy has plenty of problems that should be looked at as opportunities—in accordance with Churchill's dictum that the able person sees opportunities in every problem, rather than problems in every opportunity.

Every problem, whether technical, industrial, economic, social, or political, needs for its solution adequate information, analytical thinking as well as the willingness to take risk and accept the resulting responsibilities. Economic change usually creates more problems than one is bargaining for, and this means that dynamic industries have to be much more sensitive to market drives than decaying industries because

- they face new problems more rapidly than the others, and
- each of these problems is quite complex, with deeper implications on the life of the enterprise

Even projects that look like "a billion dollar sure thing" may capsize because there is technology risk, market risk, and management risk embedded in them. When the Tunnel under the Manche and its EuroStar high-speed train linking Paris to London were inaugurated, they looked like a grandiose achievement in the post–World War II years. A couple of decades later, however,

- the "Tunnel" was hit by a torrent of red ink, and
- technical snafus of Third World fame plagued the high-speed train

On December 19, 2009 five EuroStar trains full of passengers traveling from Paris to London remained blocked in the tunnel for 16 hours.

The reason was a thermal shock created by prevailing weather conditions. As for the passengers, they had to walk their way out of the tunnel to avoid being trapped in.

The reason given for the simultaneous electromechanical failures was that the temperature at both ends of the tunnel was subzero, while in the tunnel it is maintained at 25°C. This created a short circuit at the level of the motors. The engineers who designed this highly costly technical marvel could have done better by experimenting on likely even though improbable failures at the drafting board.

Neither every reason for the snafu was in engineering design. Critics said that a more profound reason was that the equipment had aged and its maintenance was not perfect. The snow and major difference in temperature did the rest, but such events are foreseeable. They sometimes happen, which evidently calls for much greater attention paid to reliability.

As it were, with the first 24-hour stop in rail traffic 24,000 would-be passengers were stranded in the train stations, which talks volumes about how much the alternative transport means, to the predominant airplane and ferry traffic between France and England, which can be open to malfunctions. In the end, by the time the train traffic stoppages were done with, a total of 75,000 travelers were stranded.

Trains are a nineteenth-century invention. But long tunnels with extreme differences in temperatures are a recent advance which, as this case documents, can have unexpected consequences. Technology risk is ever present, and when we talk of ROI technology risk has to be accounted for as one of the costs.

A different way of making this statement is that to properly manage research we must be artists in locating *our* technology but also our technology's risk at any given time. We must also excel in the ability to track the changing return on our R&D dollar over time—based on the relationship between resources coming in and performance being delivered.

Another important characteristic of the work we have to do as part of the investment for staying in business is to know the difference between great products and "me too" designs with meager or no future. Still another is the ability to appreciate the difference in dynamics between

- markets where diffusion is determined by the characteristics of the product, and
- those where it is determined by the size of the installed base and its absorption capability

In addition, as products diffuse across markets over time, product planners (Chapter 12) should appreciate *a monte* the groups of potential customers who need different bundles of attributes. This is part of diffusion dynamics as well as of carefully studied plan for return on investment in R&D.

Much of what constitutes a product's market success has to do with taking early on account of such differences in developing our technology and product strategy. Is our market driven by different types of end-users, or by a steadily improving technology? Is the name of the game that we have to keep evolving the technology to meet new customer segments. Or, is it that we have to do so to hit one critical application (or group of applications) better and better?

These are basic queries that practically show up with all products, particularly those that have a significant technological content. *If* their study is skipped by lightly, or the answers they are provided with are half-baked, *then* it is better to forget altogether about bringing to the market strategic products and of obtaining a decent return on money invested in research.

## 11.4 PLANNING FOR INNOVATION

The specific domains of R&D projects and their objectives vary widely, but the crucial questions to be asked in regard to budgeting and ROI are nearly the same: How much money should we spend on basic research? Applied research? Development? Can we depend on government support for basic research?

- How can the R&D budget be stretched to get better returns for the company?
- How can we evaluate at an early stage a successful research investment from an unsuccessful one?

Factual and documented answers to these queries don't come easily. Major and minor design reviews are a good "early warning" system, but they are not fail-safe. One of the best guarantees is management's courage to kill projects that have failed to meet their objectives, and its ability to take a long, hard look throughout the whole research establishment:

- The timetables
- The way budgets are used
- The rate of progress
- The results obtained in terms of deliverables

Project control is an indispensable part of planning for innovation and it should be a steady effort with well-orchestrated design reviews. Sporadic measures cannot successfully solve the problems facing a company in connection to product evolution and innovation. To survive competition in a changing market and develop a carefully studied product strategy, management must answer plenty of soul-searching questions and develop a clear policy for new products, including the cultural change this requires.

Senior management should not delegate the responsibility of an innovation policy, but learn to deal more effectively with the internal politics of innovation decision making—overcoming organizational, and information barriers. Pertinent queries are as follows:

- What does it take to create a successful innovation portfolio?
- How can we improve our batting average for choosing innovations that are likely to succeed?

Board members and the CEO and his immediate assistants must also concentrate on managing the risk inherent in testing new innovation concepts—particularly regarding those that are hard and disruptive. Proactive risk management requires skill because many of the challenges that come up by reinventing the company's product line, and eventually the whole entity, are not predictable.

Challenges connected to innovation cannot be successfully confronted without hefty changes, and companies are resistant to change. Hence the question, "How can *we* implement successful innovation practices in the face of internal opposition?" In general lines the answer is by explaining that without innovation the company will lose its market and die out.

An additional responsibility of senior management is protecting early-stage developments from being dismissed by the organization because they appear to be too risky, or too far from the firm's mainstream. In this connection, the best approach is to ask

- which are *our* alternatives, and
- do I wish to add *our* company's name to the roster of venerable firms—with names like Digital, Wang, and the old IBM

Training in industrial strategy helps. Companies that know how to survive are typically proactive. They are anticipating both articulated and unarticulated customer needs, and get themselves ready for unconventional product evolution or changing technological landscapes.

Entities that have been successful in this process learned how to integrate competitive technical and business intelligence with their decision processes. Business intelligence is instrumental in convincing managers that they should not fall in love with their current strategic thinking.

The right kind of business intelligence also helps in avoiding pricing failures. SAP's attempt to offer software as an online service, called BusinessByDesign, was a flop. It is said that a major element contributing to this has been the firm's decision to raise maintenance fees customers have to pay to get upgrades, which proved to be wildly unpopular.*

Case studies with successes and failures associated to innovative products can also be instrumental in opening peoples' mind. For instance, why is it that the Toshiba/Time Warner video recording standard DCD has been so successful, whereas the alternative solutions advanced by Sony and Philips faltered.

An analysis made by *Business Week* in early 1995, when that question was posed, came up with plenty of food for thought. At the top of the list stood four points characterizing the winners and other reasons haunting the losers.† Toshiba and Time Warner did the right choice to

- push the technology envelope to 10 billion bytes of storage,
- woo other electronics companies with an open licensing approach,
- prioritize films and games over computer applications, and
- court Hollywood aggressively through an ad hoc advisory group

* *The Economist*, February 13, 2010.

† *Business Week*, February 20, 1995.

By contrast, Sony and Philips followed the wrong strategy. The three main failures have been sacrificing capacity for compatibility with existing disks; slowing the development of a high-capacity digital videodisk prototype; and presenting Hollywood with finished specs, instead of seeking advice.

It is appropriate to notice that while these choices ended in sorrow for Sony and Philips, not all of them were outright wrong. For instance, as far as compatibility is concerned, Intel faced exactly the same challenge with its microprocessors. For long years, the strategy of upward compatibility worked right for Intel, but it was the wrong choice for Sony and Philips.

As this example helps in documenting, the choice of R&D goals and projects set up to serve them determines the fundamental nature of a business—from its future prospects to its relation with its customers, competitors, suppliers, and personnel. The product line establishes the limits of opportunity for expanding a company's horizon, which can be no bigger and not much more profitable than the products in which we compete.

- Product targets are at the heart of strategic planning, and therefore of defining R&D projects.
- They are the starting point for an overall corporate master plan since they determine capital, personnel, and facility requirements.

In planning for innovation, coordination between R&D, marketing, and finance is a "must," but none of the three divisions should have a *veto*. To be effective in the longer run, such coordination must be a culture characterizing the relation between the company's divisions.

## 11.5 DON'T SELL QUALITY TO BUY MARKET SHARE: TOYOTA'S FAILURE

Marketing should evidently enter into planning for innovation, and therefore into product decisions. This is one of its key missions. But it is wrong to overweight marketing goals at the expense of technology and of quality, as Toyota's experience in 2009 and 2010 documents.

To appreciate the depth of the precipice to which Toyota threw itself by putting ego and market share above technological excellence (which was for decades its competitive advantage), we should briefly return to end of World War II—and Japan's effort to come up from under. The man who contributed a great lot to the rebirth of the Japanese industry on solid quality foundations was W. Edwards Deming, an American engineer.

Dr. Deming had a passion for creating a system that could be trusted to make better products. His method for doing so was rigorous quality control, which the industry of victorious America thought that it was something of a luxury, and told him so.

Not being accepted as a prophet in his own country, Deming joined General McArthur's staff in Japan. He was a person who listened as much as he talked, and hated waste. He also felt that the U.S. industry had become wasteful; to the contrary Japan could not afford that because it was not abundant in natural resources.

Scarcity of resources evidently played a role in the decision of the Japanese industry's captains to follow Deming and his quality bible. He taught the Japanese that quality was no minor function that could be accomplished by asking the workers at factory floor to be more watchful. *High quality demanded total commitment by management.*

- *If* the CEO and the senior VPs were not committed to the idea of quality, and
- *if* executive promotions were not tied to quality,
- *then* the quality of the company's products will be dismal.

Deming's bet has been that by convincing top management that quality is king, while quality control's standards and methods are sacrosanct, a quality spirit will be developed and seep down the organization into the middle and lower management levels; therefore, inevitably, to the workers. By contrast, *if* the company treated quality as an afterthought, *then* true quality would be a chimera.

It helped that the Japanese were by all evidence aware of the low level of their manufacturing quality—both before the war when Japanese goods were known for shabbiness, and during the war when Japanese engineers saw American weapons systems consistently outperform their own. Therefore, they were eager to learn from American engineering and manufacturing skills.

In addition, W. Edwards Deming had a method for steadily improving the quality of industrial products by applying statistical quality control (SQC). Credit for this goes to Walter Shewhart, a physicist who in the

1920s and 1930s, at Bell Laboratories, had pioneered in the use of statistics and statistical charts aimed to assure higher industrial quality.

It is interesting to take notice that during World War II the American military successfully used Shewhart's techniques. But after the war ended U.S. industry lost interest in them. (I had myself learned SQC during postgraduate studies at UCLA and successfully applied them both in the United States and in Europe. At Osram, the German lamp company, SQC improved outgoing quality almost by order of magnitude.)

Deming had learned SQC at Stanford University, and his skills were greatly in demand as long as the War Department demanded that many of its defense contractors apply Shewhart's methods. But all that suddenly changed with V-Day, the mushrooming affluence of the postwar years and the (wrong) feeling that no challenger could come up from under.

A few decades down the line, starting in the late 1970s, the American industry lived to regret the day that it had lost its interest in SQC. The generation of managers who felt there was no need, no money, and no time, because the company seemed able to produce "the old" way, started in its way to retirement and the people, that replaced it lacked SQC skills in America—though they were abundant in Japan.

To make his point about SQC's vital importance, Deming liked to tell about one engineer who had worked for Western Electric during the war, when it had been a center for quality control and when the Western Electric Control Book was a kind of quality-control bible. That man left Western Electric in the late forties and returned 8 years later to find that all the control charts had simply disappeared.*

In Japan, it was different. In 1946 and 1948 while in Tokyo Deming met some Japanese engineers who knew of Shewhart's seminal work. They had even translated Shewhart's book, but they were unsure on how to apply SQC. The soil was fertile for transplanting Shewhart's ideas.

Deming was asked by the Japanese in 1950 to lecture on quality control. But reportedly he suspected he was going to be wasting his time till Ichiro Ichikawa, one of his sponsors, was instrumental in putting together an audience of 45 top Japanese industrialists. To keep their spirits up, Deming told them that if they listened to him they would be competitive with the West in five years. To his listeners it seemed impossible, but by the mid-1950s it was done.

---

* David Halberstam, *The Reckoning*, William Morrow, New York, 1986.

The Japanese love affair with quality did not last forever. By the 1980s in Japan, like in America after World War II, other issues took the high ground, while product dependability was relegated to lower status. That was the decade when Japanese industry swept the globe with its products and Japanese banks thought they can conquer the world but failed—taking down with them the country's whole economy.

The 1990s in Japan have been the so-called lost decade that has never really ended. Great companies like Sony lost their pizzazz, but Toyota labored on and in the early years of this century the second raters who run the company's fortunes targeted the No. 1 position in car manufacturing rather than customer satisfaction. Quality, which should have been the flagship, was nowhere in the agenda. Instead, the headline was

- sales, sales, sales, and
- some dubious "innovations" like electric cars (a century-old product) and hybrids

Toyota is by no means the only company where product planning dropped quality to the waste basket. This negatively reflects on management because good governance never slackens its watch over quality problems. Instead, they are a steady preoccupation.

Toyota failed in this score. The way an article in *The Economist* had it, there is a widespread belief within the automotive industry that Toyota is the author of most of its own misfortunes.* In his testimony to the U.S. House of Representatives Oversight Committee on February 24, 2010 its CEO, Akiro Toyoda, acknowledged that in the pursuit of growth his company stretched its lean manufacturing principles—and even gave less weight to its old motto of focusing on putting customer satisfaction above all else.

After having based its fortune on *quality*, Toyota changed priorities. By all evidence the switch came in 2002, and is considered to be the dumbest decisions its top management made.

- The conquest of global markets got the upper ground, with research, development, and quality becoming subservient.
- Having abandoned the "quality first" principle, Toyota started working with a lot of unfamiliar suppliers who did not have the credentials to uphold the company's past culture.

* *The Economist*, February 27, 2010.

Nothing was learned from Intel's stumble in product quality when a minor flaw was discovered in the Pentium that affected some mathematical calculations. Rather than rush to correct the problem, Intel tried to downplay it—a strategy that quickly turned into a public-relations disaster. (Eventually, Intel was forced to offer a replacement for all affected chips, at a cost of nearly half a billion dollars.)

Intel learned its lessons, but other companies did not take due notice. Dr. Andrew Grove now thinks this episode actually benefited the firm in two ways. It proved to internal skeptics that microprocessors really had become a consumer brand. And it bolstered his efforts to *improve the quality* of manufacturing, to protect the firm from future fiascos.*

## 11.6 SECURUM: USING TECHNOLOGY TO BUILD UP DEFENSES

Real estate loans has always been a risky business, as attested by the losses of American banks in the 1980s, Japanese banks in the early 1990s, and again by American banks in the real estate abyss of 2007–2010. In late 1993/early 1994 Germany, too, headed for its own version of a real estate bubble, and at the epicenter of it was Schneider's bankruptcy.

The collapse of the Dr. Jürgen Schneider AG, a well-known construction company, occurred at a time of stress for the German banking industry because a swarm of firms were struggling cross-industry sectors. In January 1994, bankruptcies of all kinds were up to 22.8 percent year-on-year in western Germany and 99.2 percent in eastern Germany. Dismal economic conditions did not bode well for real estate.

- At the end of 1993, in Frankfurt, prime office space was going for $44 a month per square meter, down from $55 two years earlier.
- In Leipzig, rents were 44 percent lower than in 1991, while in Berlin the decline was 24 percent.

There were as well other signs that companies like Schneider's may be heading for the precipice, even if German banks continued lending.

* *The Economist*, September 9, 2009.

The Schneider AG was a special case because it was regularly sued by subcontractors for not paying its bills.* It came, therefore, as no surprise that when in April 1994 it went under, leaving behind a huge sum in debts. But even the experts could not figure out how deep the dry hole was.

- Original estimate spoke of DM 5 billion (then $3 billion).
- Shortly thereafter it was said that the money down the drain exceeded DM 8 billion (nearly $5 billion).

After two weeks of being bashed about its handling of loans to the real estate magnate, Deutsche Bank had to admit that its management of Schneider credits was abysmal. Hilmar Kopper, the bank's chairman, and board members Georg Krupp and Ulrich Weiss, faced the press to explain the facts and the consequences.†

A subject they carefully avoided was that Deutsche Bank executives (and many other German bankers) thought that they had solid collateral against their loans to the Jürgen Schneider company. But they were unpleasantly surprised when Gerhard Walter, the liquidator, said that what will be secured is about 10 Pfennig to the Deutsch mark (say 10 cents to the dollar), and banks might lose as much as DM 4.5 billion ($2.7 billion) on the DM 5 billion they had lent.

As it often happens in such cases, an expensive miscalculation like this was not the only bad news. Germany's recession has led to a rapid increase in other insolvencies. For the 13,000 West German firms that went bust in 1993, bankruptcy was a messy business.

There was as well plenty of folklore. Information uncovered in mid-April 1994 during a joint examination of the company's records made by Schneider executives and officers of the Deutsche Bank was said to have revealed that bribes were paid to appraisers that allowed the company's owner to inflate the value of his properties.‡

A most interesting case was that of a shopping mall built just 500 m away from Deutsche Banks Frankfurt headquarters. In this the bank had allegedly sank into the coffers of the Schneider company in exceeds of

* *Business Week*, May 2, 1994.
† *Herald Tribune*, April 25, 1994.
‡ *International Herald Tribune*, April 29, 1994.

DM 1.2 billion ($720 billion) theoretically covered by real estate collateral. Practically, however,

- the values were not only inflated by square meter, but also the declared covered area itself had little relation to reality, and
- the Zeil-Galerie Les Facettes was registered as featuring 20,000 $m^2$ of an area that can be rented, but the real size of rental area was about 9000 $m^2$

"We have made errors," said Hilmar Hopper the Deutsche Bank chairman, but swiftly added that there will not necessarily be fallouts on those banks' executives taken for a ride. By contrast, a spokesman of the Deutsche Bundesbank was of the opinion that the different credits to the Schneider company did not respect the elementary rules of the laws, which regulate collateral and the extension of credit.*

Such sloppy work was surprising because all the loans officers and senior bank executives who authorized these large credits had to do was to go 500 m down the road and see by themselves if the mall being built was 20,000 $m^2$ or 9000 $m^2$—as well as its status of advancement. Or, alternatively, they could use technology to exercise management control. That is where Securum's solution comes in.

Securum has been a Swedish institution set up by the government to manage and dispose bad loans, holdings, and other unwarranted investments of bankrupt Nordbanken.† From day 1, Securum used technology to gain competitive advantage over the derelict status of Nordbanken projects. Most important to our discussion is a graphics system developed to track its real estate properties in an interactive manner.

The first component of this solution, which was most novel for its time, has been a scanning program fetching the building's drawings and registering them in a database after vectorizing them. Subsequently, a virtual reality graphics routine enabled the user to

- walk through all the drawings
- get information about the building

---

* *Le Figaro*, April 26, 1994.

† During the banking crisis which hit Sweden in 1991 Nordbanken, the country's second largest credit institution, approached bankruptcy. The Swedish government, which was also the major shareholder, came to the rescue to the tune of $10 billion. But it also took away from Nordbanken all the bad loans and all its holdings in compensation for the $10 billion, and set up Securum to manage them.

- make a floor-by-floor inspection, and
- select and make up to six layers of changes in the drawings to update them with the building project's advancement

The graphics system has been integrated with real estate accounting for online updates, upgrades, and analyses. Its functionality has been enriched by an expert system that assists the user in selecting the changes or repairs to be done in any place in the building; still another expert system provides the complete specification of the job. Securum's solutions also provided means for

- budget control, and
- plan versus actual analysis

Securum was able to capitalize on technology because its IT was run by young dedicated people with modern ideas. Working through rapid prototyping on client servers and applying a fast deployment policy, after one year they had in place not only a complete banking system but also the aforementioned real estate administration aggregate and plenty of other sophisticated applications.

Securum's solution was innovation at its best. The results obtained through information technology demonstrated what a small but capable and fast moving group of IT specialists can provide versus what ossified data processors are doing on huge expensive mainframes. Plenty of other examples can be given to support this reference, one of the best being the aforementioned Jürgen Schneider scandal at Deutsche Bank whose control was based on

- loans officers doing a substandard job, and
- old technology incapable of providing analytics

The Deutsche Bank mainframers were unable to control the risks involved in the Schneider loans and this cost the financial institution, for which they were working, an estimated DM 5 billion ($3 billion). By contrast, Securum did a superb job at very low cost and very rapid development timetable. A high degree of flexibility, novelty of ideas, and fast implementation make the difference between new and old IT—and at the end between profits and losses.

## 11.7 THE RIGHT FEEDBACK ON PRODUCT INFORMATION

Feedback information on new products is very important and its evaluation must be done rationally to avoid losing time, increasing costs, or blurring the issue. A substantial part of the problem in assigning evaluation responsibilities has to do with possible conflicts of interest by those expected to perform this task.

Experience teaches that it is not enough to have a general plan about the analysis of monolithic feedbacks. If we are going to spend R&D money to develop products and marketing money to promote them, we need to define successive steps in judging obtained results. There is no way of developing an effective innovation policy without intermediate goals—each one leading to the next.

Komatsu, the Japanese company that makes construction machines such as excavators, bulldozers, dump trucks, and wheel loaders (as well as industrial machines and robots) applied this principle. When it was still a relatively small company, its management developed the strategy, "*encircle Caterpillar*" and required that feedback information is based on six pillars:

- Quality
- Product range
- Value differentiation
- Dealer network
- Cost swamping
- Competitive pricing

Because Komatsu was export oriented what mattered the most in pricing is how the firm's products held in dollar terms. At the time that decision was reached, the yen/dollar exchange rate was 180, giving Komatsu a huge price advantage. But then the yen rose significantly. When it crossed the 100 to the dollar exchange rate Komatsu repositioned itself and its products to be able to compete with an exchange rate of 80 yen to the dollar.

Notice, however, that price was the sixth pillar; not the first. Above every other criterion Komatsu bet on quality first. As Section 11.5 brought to the reader's attention, W. Edwards Deming had brought to Japan this culture. He also instilled through his teaching the principle that the company's engineers should not be separated from the manufacturing line, housed in some nice sanitary building. Instead, they should

- be out on the factory floor as much as possible, and
- be able to observe and record how the line workers act in quality terms, as well as what they suggested in improving quality

"Quality first" is not the principle one hears the most often in industry—as quality is being made subservient to cost. In the longer term, this proves to be a disastrous policy as Toyota and a long list of other companies learned at their expense.

A great deal of lessons can indeed be learned in terms of the importance of quality assurance from Richard Feynman's findings in connection to his investigation of the Columbia disaster. This is one of the best examples that can be found at the junction of product information and strategy—whether the product is designed for space travel, the military, manufacturing, or banking.

1. The shuttle's budget was decreased by 40 percent because the spaceship was considered a mature product.
2. There has been life cycle uncertainty, the shuttle's replacement originally projected for 2005–2007. It slid to 2012–2015, and to beyond 2020.
3. The maintenance and test equipment was 22 years old, way beyond state of the art.
4. In terms of feedback, poor data quality played a key role to the disaster keeping reasons for incoming failure opaque.
5. With poor data sets and flawed judgment, came overdependence on simulations—remote from the salt of the Earth.

The sixth finding by Feynman was perhaps the most important. He pointed out that mismanagement, too, was present at the space agency. Managers had no understanding of how their organization worked, and were lulled into complacency by the shuttle's successes.*

It has happened with NASA, and it has happened everywhere.

Remoteness from the salient problems, inability to challenge the "obvious," little or no attention to quality and fuzzy data on which decisions are based, are precisely the reasons that bring a person, a company, an agency, and a nation in distress.

---

* *USA Today*, August 27, 2003.

# 12

## Product Planning and Pricing

### 12.1 THE PRODUCT PLANNER

*Product planning* and *profit planning* are two functions characterized by a fair amount of integration, and this not only because the latter depends on the former. Product planning is directed toward an analytical and rational evaluation of the market, product line, drive for innovation, and competition. Because it involves both financial and technological decisions, it tends to merge with business planning—where profit planning is cornerstone.

- The return on investments in R&D and engineering will depend on new products, their market appeal, and their pricing.
- New product choices, their development, management, and follow-up in the market is the *product planner's* job.

To be regarded as functional executive and advisor to top management on matters concerning the product's future, the product planner is first and foremost responsible for solving product initiation, evaluation, and evolution problems. He or she should be able to demonstrate a sound judgment as to how the company products compare with the competition:

- What kind of products will customers and businesses be asking for?
- Is *our* company's R&D able to develop such products?
- What quantities of projected product(s) can be sold, at what price(s), and over what time period(s)?
- Are projected timetables acceptable or will competitors decream the market?
- What kind of competition will *our* company's products face in the years to come?

Answers to the latter three questions should be given jointly by product planning and marketing, but the product planner must also handhold with R&D. Able solutions to product planning must have globality, benefit from technology, and be subject to crucial profit-and-loss (P&L) evaluations, as Figure 12.1 suggests.

Typically, the product planner bases a fair amount of his work on checklist data derived from marketing analysis, technical planning, and research results. He draws up time plans encompassing all stages of development and production for each product in his portfolio. Through a timely, coordinated action, he aims to diminish the margin for error in planning for the future. The results will be largely felt in the company's profit figures. In a fast-moving marketplace, no firm should underestimate the value of being better organized and quicker to respond than its competitors.

At the core of the activities in Figure 12.1 is *innovation*. According to François Michelin, who has been the spirit behind the No. 1 manufacturer in the world—and who served the company from the age of 29 to 83 as its CEO and the guide of the management who succeeded him,

- An innovation must serve to enlarge the market.

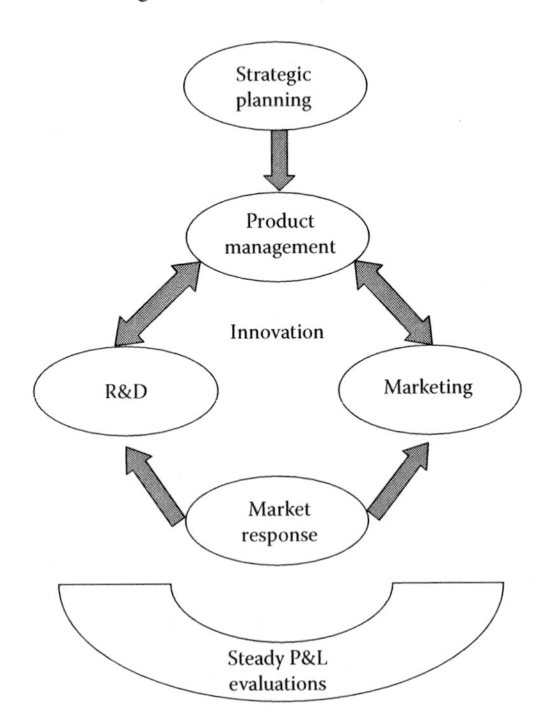


**FIGURE 12.1**
Able solutions supporting corporate strategy must have globality, benefit from technology and marketing, and be subject to critical profit-and-loss evaluations.

Theoretically, this is so, but practically it is not because many CEOs who may be generally good managers don't have the salt of the earth of the company they run and of its products. Therefore, Michelin insists that

- a CEO must learn his art through all levels of the company, not just by residing at the top.

Innovation is the result of trial and error, said Paul Galvin who invented the portable telephone and made Motorola a global firm. Therefore success may come after a series of errors. It follows that the CEO who does not know intimately his firm and its products is at disadvantage in performing this task since he is not well placed to appreciate all factors contributing to a stellar product.

Plenty of other opinions have been voiced by successful businessmen about innovation. One of them comes from Akio Morita, of Sony, who had three management principles. These should be written in block letters and laid on the desk of every executive.

- Innovating without perfectly adapting the product to the market, serves nothing.
- Only the ambitious people who care to investigate should be hired.
- We all do mistakes, but we should not make the same mistake twice.

A different viewpoint was expressed by Robert Woodruff of Coca-Cola: "To sell you must first create in the client's mind the desire for the product. Innovation and marketing correlate." To this, Alfred Sloan was to add, "The style and price of products must flatter the ego of the clients."

Quite similarly, a deep knowledge of the company and its products is required from product planners, who must work within the realm of their company's strategic plan, starting their work with the concretization of the customers' and the market's wishes and continuing with follow-ups on the product's R&D—regarding its innovation, functionality, quality, marketability, cost control, and timing of deliverables.

- Chief executives play their company's future through innovation.
- The product planners' contribution is the integration of technical, financial, and sales experiences.

The responsibility of the product planner shows in setting the parameters of a new product; developing and applying experimental approaches to marketing analysis and product technology; making measurements that are inherent to the development of the preliminary specifications; and following up the work going on in prototype construction and manufacturing.

He is responsible for preparing and publishing data and specifications, generally material that describes the aspects of his work, including authoritative conclusions on the problem under investigation. By bringing into focus the key factors in a new product problem, product planning must examine

- technical contents
- foreseeable market performance
- financial implications, and
- coordination issue involving different company departments

Product planning acts as a central authority accepting inputs from different areas and providing feedback information. The market appraisal for a projected new product must be made by a joint effort between engineering and marketing with the object to arrive at a preliminary functional specification. Not all product planning effort is, however, well done. Two of these reasons are as follows:

- Product life cycles tend to be shorter.
- Specialization tends to narrow product appeal.

To perform his mission in an able manner, the planner should increase the likelihood of avoiding past mistakes, ruling out the subjective element in new product decisions. A major part of his assets will be his ability to compare resource requirements to projected business opportunities and expected benefits. To do so in an able manner he needs to

- realistically analyze the company's R&D skill and deliverables history,
- relate the firm's capabilities to the market's drives and size,
- evaluate the product's likely performance against estimated market demand,

- examine product price-profit interrelationships to decide about the product's potential, and
- submit his conclusion to top management based on a consolidated statement on product-market opportunities

Sometimes, the excitement associated to a new product-in-the-making is so high that it biases the experts' opinion about its potential market size. At the beginning of this century, pundits predicted that sensor technology and online input will be revolutionized with *smart dust*. While there exist some smart dust applications, particularly military, its supposedly great potential has not yet been realized.

But there are companies working on it. In 2009, Hewlett-Packard began a project called Central Nervous System for the Earth, a 10-year initiative to embed as many as a trillion pushpin-size sensors around the globe. Combining electronics and nanotechnology expertise, its researchers aim to develop a system of sensors with accelerometers that are as much as 1,000 times more sensitive than the commercial motion detectors used in Nintendo Wii video game controllers and some smartphones.* The practical use of all this is a different matter.

Though it is not very often talked about, another basic function in the product planner's remit is to evaluate the death of products in the company's pipeline, which would allow a timely and orderly search for replacements. The death of products may come for one of several factors:

- Obsolescence
- Loss of market appeal
- Uncompetitiveness, because of functional or cost reasons
- Expiration of patent protection, and more

When products die, they may bring down the drain the company making them. Prior to being bought by Warren Buffett and converted into an investment powerhouse, Berkshire Hathaway was a textile firm fading away. One time, in the late 1960s, a Berkshire salesman was with a customer on Fifth Avenue trying to sell handkerchief cloth. The customer pointed out his window to the women going into Lord & Taylor and said, "You see all those women carrying pocketbooks? There is a box of Kleenex in every one. And that's the end of the handkerchief business."

* In November 2009, HP announced that it had a prototype.

## 12.2 PRODUCT PLANNING AND BUSINESS OPPORTUNITY

The more diversified the company is, the broader becomes the range of planning activities. These can be described in three different time zones: immediate, up to one year, and at five or more years.

Product plans must reflect business opportunity and, as such, they have an impact on market strategy. Fundamental to a product planner's, or a product planning committee's work, is the accurate analysis of the market. It includes the perceived needs of customers both now and in the future, but also best timing of new announcements—which are often conditioned by technological breakthroughs.

A very critical question only the better managed companies ask themselves is, "Can our R&D deliver what this project idea needs to become reality?" When properly conceived, an engineering project brings together investigative, creative, design, and testing activities. The organization of men, materials, and facilities toward a desirable end is an almost self-contained management process.

For these reasons, the engineering project forms a natural unit for which responsibility can be assigned, over which authority can be given, and the results of which can be measured. These are exactly the conditions needed for the application of planning and control techniques. But company politics bias what would have been otherwise a neat technical project.

As consultant to the board of AEG-Telefunken, I was often asked to perform analyses of ongoing R&D projects and of their ability to meet product planning timetables. Sometimes the results were positive, and everyone at board level was elated. But in many cases they were negative—and then company politics went into action.

One case I can talk about concerned a never ending project of a new magnetic tape reading station, which had overrun its budget and timetable, while still being plugged by technical problems. My advice was "kill it." The director (SVP) of engineering complained to the president that he was unfairly targeted, and the board decided to continue spending money for nothing.

Another nonconfidential case regarded a new "Selectric-type" typewriter for Olympiua, the company's business equipment unit. The concept was good, but at a time of office automation (early 1970s) the market for selectrics was fading. "Drop that project," I suggested. It took a whole year to decide that this was indeed the right thing to do.

Both these two examples and a source of dollars had violated what a business opportunity analysis is all about, and most particularly the fact that its hypotheses and statements are time sensitive. Though they may have been tested through market research, as time goes by they become obsolete.

After they have been made, business opportunity statements are being challenged by market changes; they don't last forever. Statistics I kept over the last 30 years indicate that eight out of ten new products have been business failures because

1. they were too often evaluated apart from its environment,
2. timing was neglected, or irrationally extended,
3. too little consideration was given to competitors and their products,
4. assumptions made at the beginning of product planning were never challenged afterward,
5. the skills and capabilities of salesmen who must sell the products were ignored, or taken for granted,
6. the salesmen were not properly trained in new product functionality and possible headwinds,
7. promotional campaigns were conceived in a vacuum, based on the company's introverted views, instead on market drivers,
8. the company continued with the product even after failure was evident,
9. management failed to eliminate aging products to prune the product line and to allow new product to grow in the market, and
10. the marketing budget was meager compared to R&D outlays (As stated, the case should be exactly the inverse.)

In addition, business opportunity needs proper financing. But throwing money at the problem does not help. Too much money would not assure success, and will probably lead to spoilage. Therefore, the board and CEO must give careful consideration to one of key aspects of product planning activities: *budgeting.* Clear-cut lines of functional contribution contribute a great deal to an objective allocation of funds.

As Section 12.5 will bring to the reader's attention, market appeal, pricing, and profits will be better protected when a company structures its product line to cover all segments of the market to which it appeals and avoid product overlaps. Though products line development is to a significant degree a function of technology, in its fundamentals it is primarily related to

- correct identification of the market segments, and
- the tagging of products at prices the market can afford

This is evidently part of a pricing strategy aimed to maximize profits within the planning horizon targeted by management decisions. What's the goal we are after? Long term? Short term? What's the price of continuity and good will? Are we ready to face competitors' actions like price wars? (See Chapter 13.)

To give an adequate assessment, management must use its broad understanding of the company market and product planning process—which are continuous processes, though planned results are formally stated only at prescribed intervals. Typically, such intervals correspond to those of financial allocations and give rise to *plan versus actual* comparisons.

Chapter 14 will explain why the budget, as annual financial plan, does not constitute an automatic authorization to spend money. It makes an a priori allocation of financial resources but leaves it up to management to authorize expenditures—which must be done

- with care, and
- in full appreciation of the fact that costs matter

Financial analyst must bring a new product cost figures into shape by asking pointing questions and challenging assumptions. Product planning must offer nothing short of a well-done business opportunity report that

- identifies factors related to economic value problems,
- emphasizes matters influencing the possible profit-and-loss behavior of the product once in the marketplace, and
- assists in predicting P&L—which in turn indicates *if* the product should be introduced at all

Income forecasts should be based on market potential, pricing, and length of this product's life cycle. A business opportunity must be budgeted. Projected cash flow must make feasible a certain degree of assurance relative to the cash position of the firm if the company makes a financial effort.

The proper exercise of management control (Chapter 7) requires that the product plan provides feedback for revaluation in terms of income-related projections. This calls into play budgetary chapters—and no budget can be considered worthwhile unless past discipline shows that it is worth its salt. This has to be attested through comparisons between budgeted and actual figures.

## 12.3 NEW PRODUCT PLANNING METHODOLOGY: A PRACTICAL EXAMPLE

Section 12.2 made the point that the product planning process can be drastically improved if—while new projects are being drawn up and studied—the possible financial consequences can be determined reliably. This may sound obvious, but many companies don't put together the skills necessary to do such studies.

This is regrettable, because there is plenty of evidence that product and profit planning turns the long-range plan into an action program with significant financial overtones. In addition, the often discarded factor of timing impacts on the framework of "joint effects" that become visible in the medium to longer range.

As Table 12.1 brings to the reader's attention, the activities involved in the evaluation of a new product idea take time. This should be scheduled in a way allowing for product plans to be checked against other business plans and, modified to produce a cohesive and realistic total.

Another basic ingredient is a good methodology; the subject to this section. Tables 12.2 and 12.3 define the way in which a leading

**TABLE 12.1**

Evaluation of a New Product Idea

| Identification | Evaluation |
|---|---|
| Product idea | Examine if new product idea falls within the company's objectives |
| Marketing and competitive research | Estimate customer response through market research and competitive performance |
| Business opportunity report | Define needed integration of research, engineering, manufacturing, marketing, and financial factors |
| P&L | Estimate most likely profit and loss |

**TABLE 12.2**

Technical Product Definition

| Identification | Planning Job |
|---|---|
| First system analysis | Determine functional specifications, including matters relating to initial design, cost, and price estimates |
| First engineering report | Prepare a preliminary engineering proposal and a general statement on probable schedules and costs. Include all other pertinent factors to the design under study |
| Original definition report | Formulate an original definition of overall product plans, including their timing. Consider engineering, manufacturing, marketing, and financial data |
| First profit-and-loss report | Integrate estimated costs with projected "market potential times price," and formulate a profit-and-loss approximation for the product in question |
| Product program authorization (PPA) | Integrate tentative definitions and estimates of the plan for a proposed product program to provide a well-thought-out basis for decision. |
| | Obtain the chief executive's authorization to establish (modify or discontinue) a product program |
| Basic software development | Define the software characteristics, development time, development cost, standards, and criteria for "do or purchase" |
| Testing procedures and quality assurance | Identify the step-by-step testing procedure, including destructive testing (if necessary), statistical quality control (SQC), responsibilities for the testing programs and for accept/reject decisions |
| Reliability report | Proceed with a mathematical reliability study. Use simulation methods to "construct" hardware systems. Consider the human factor in man-machine interactions. Arrive at statistically significant reliability results |
| First marketing plan | Following the PPA, work out the Field Maintenance Service (when applicable) and other service requirements. Also the means of meeting them. Maintenance should analyze the hardware to improve field work methods |
| Applications support | Establish the nature, extent, and cost for applications support action. If the development of any kind of software is necessary, decide whether it will be done internally or externally, define its nature and extent, estimate delays, and get approval for budgets |
| First competitive evaluation | In collaboration with marketing, develop concrete estimates on where the competition will stand in 1, 2, 5 or more years for the product in question. Decide, at least in a preliminary way, on possible competitive strategies |

**TABLE 12.3**

Product Program Development

| Identification | Planning Job |
|---|---|
| Engineering/manufacturing specification | Initiate work in engineering specifications, and manufacturing engineering study for process development |
| Appraisal report | Conduct appraisal studies for both product and manufacturing engineering, along with realistic cost estimates for work to be accomplished |
| Manufacturing order | Initiate the manufacturing work (first series) with the objective of timely and quality deliveries within budgetary limits |
| Final marketing plan | Formulate detailed marketing plans. Launch information and demonstration sessions. Work with advertising in building momentum for product support |
| Information plan | Develop informative literature, and user instruction manuals. Work out all assistance requirements. Develop maintenance instructions |
| Final service plan | Establish a definite schedule for retraining the field support force (if applicable). Study the organization and administration of the training program |
| Final analysis of competition | Formulate a final competitive intelligence and analysis report. Show probable strategic moves by competition and the company's ways and means for counteracting them |
| Total progress report | Refine the final planning process by critically evaluating performance against the desired new product ends. Feed data back into the continuous planning process, and after through analysis, report to top management |

manufacturer has handled its product planning activities within each phase of a new product program. (This firm's product line is information technology.)

A product screening committee may be assigned responsibility for the preliminary evaluation of new product ideas, giving top management the results of its evaluation. Among key questions this committee should answer are as follows:

- Who *is* the customer?
- What's our opportunity?

- How many resources will be committed and for how long?
- What is the financial reward at the end of the day?

The committee may also ask R&D for a full-scale laboratory investigation or a breadboard model; as well as manufacturing engineering for production process specifications.

***

The work briefly described in Table 12.1 covers the initial basic steps for investigating a product idea and establishing whether it looks attractive. Results may show that a certain product concept should not be examined any further. For instance, investment along this line may be unwise because of low cost/effectiveness. Alternatively, a business opportunity may be positively assessed, in which case the activities described in Table 12.2 are initiated.

The technical problem definition targets factors related to economic value problems emphasizing, among other issues, matters that influence the profit and loss of the product. Beyond what Table 12.2 brings to the reader's attention, pertinent information includes competitive analysis studies, economic research, customer discussions, records, as well as estimates of sales and service.

The work described in Table 12.2 should fulfill not only technical but also economic and financial requirements for a proposed product.* To a considerable extent the study must be customer centered including the value of a new product to the user, evaluating competitive products and prices, and estimating the maximum obtainable price during a specific period of time.

Similarly, it is important to determine possible conflicts with other commitments that have been made and are still in the pipeline. This is one of the ways through which risk associated to overlaps and incompatibilities may be contained—though a certain amount of risk always exists with timing, costs, and the failure to anticipate the competitors' moves.

An integrative view of the whole process is presented in Table 12.3. The activities it identifies assist in developing a product program in accordance with top management authorization; updating the tactical and strategic plans; and revamping portions of the new product development process that require more precise definition.

* The procedure in Table 12.2 allows an orderly formulation of product definition, but has the disadvantage of crowding the product planning department with work. This can be solved by outsourcing some of the functions to other departments.

By dividing the product planning function into specific stages, management's attention focuses on achieving the most cost/effective solution. Step-by-step, factors working against product profitability must be properly identified. Good profit figures do not come around accidentally.

Along with the financial estimates and evaluations, it is important to develop accurate time estimates. A sound time plan should integrate overall departmental and divisional agreements on the work program to be accomplished:

- Accounting for available human, material, and financial resources
- Reflecting prevailing constraints including those associated to procurement
- Anticipating timing problems and bottlenecks

Procurement is a steady challenge to every company. Its optimization requires a substantial search of the market for procurement sources and involves not only continuity but as well cost/effectiveness and reliability considerations.

The proper procurement study culminates in the release by product planning of a "procurement policy" paper, which, after obtaining the CEO's approval, is sent to the company's procuring agency. Product planning coordinates this until the completion of the first major procurement.

By going through this formalized procedure, whether coordinated through a department or a committee, product planning will go through a number of strategic decisions. A good deal of them (but not all) will be concerned with *product issues*:

- The evolution of each product line (growth, decline)
- New product needs (or product improvements)
- Trends by product and replacements
- Extent of market coverage by product (and product line)
- Competitive product evaluations (and substitutes)
- Performance of products and services in terms of customer needs
- Reactions to products and services by customers (and by the sales force)
- Observance of quality standards and specifications
- Guarantees, warranties, after-sales service
- Profit margin by product (or service) and by product line

Although no universal rules exist in product planning practice, studying what has been done by successful industrial firms helps in identifying what should be done, and what should be avoided. Most of the factors described here are instrumental to ensure that the product plan is able to screen the proposed product's ability to meet economic criteria; planning for profits is and should remain a primary goal in any product planning task—which leads us to the importance of factual and documented product pricing.

## 12.4 PRODUCT PRICING THROUGH REVERSE ENGINEERING

Product pricing integrates into itself the cost of staying in business and cost of goods sold, including marketing and administrative expenses. At the same time, product pricing is a function of market demand and of competition. In classical economics, prices are typically established on the base of one's own cost of production (direct labor, direct materials depreciation, amortization, overhead) and distribution. Today this process has been inverted. Prices must be established while

- accounting for globalization, hence the costs of other vendors no matter where they are located,
- integrating into pricing decisions the risks the company is assuming, including much lower costs, risks, and
- considering future events, which themselves are uncertain, as well as opaque in a growing number of cases

As a result, pricing increasingly resembles the nonlinear process employed in insurance. The most critical factor in this nonlinearity is *uncertainty* of an outcome, which while inherent in every business, has grown in importance to the point that it became a cornerstone notion of the service economy.

Books say that pricing decisions must be conditioned by a profit margin evaluation, which have evident effects on the company's financials. That's true in a theoretical sense. But practically with the exception of the company which has a unique product and therefore no competition, every other firm sees its price list turned on its head by the global market.

The time when a company wrote its price list after accounting for costs and profit margins is long past:

- Today the price list is written by the market, and
- every company has to meet it by innovating, and has swamping internal costs

There exist some stratagems that over a limited time permit to escape from the market's pricing straitjacket. Ford Motor Company's Mustang, in the 1960s, was not designed to be a sports car abiding by the preconceptions of what a sports car "should be." Lee Iaccoca, who led the Mustang project, had discovered that a market segment existed that valued sportiness in a car but was unwilling to pay the price for a sports car. The task Ford successfully completed was to design a car sufficiently sporty to satisfy this segment but stripped of those elements of a sports car that could drive its costs up.

This is one of the best examples of a product pricing strategy that works by *reverse* engineering—it is focused, customer oriented, and able to deliver results. With buyer-centered pricing, Ford anticipated a price even before the product's development, and provided evidence of what can be achieved through reverse engineering the *price tag*—which becomes a price target.

This is a totally different ballgame than classical pricing procedures that can be found in textbooks. Dynamically upkept tentative price tags are a very important guide to product development, since they mark potential products as candidates for acceptance, rejection, or thorough redesign.

By recognizing early enough those products for which tentative price tags are too low or too high, a company gains significantly more freedom than the legacy pricing methods allow. One of the challenges of buyer-oriented pricing is to identify different market segments and design a strategy that effectively distinguishes important client groups attracted by "right pricing." Airlines, for example, successfully distinguish segments simply by requiring a long lead time to reserve a low price fare.

- Vacationers are very price sensitive but also easily able to anticipate their demand for air travel.
- By contrast, business travelers are much less price sensitive, but require flexibility and immediate response to changes in their schedules.

By offering lower fares with inflexible scheduling, airlines can attract enough price-sensitive travelers to support the frequent flights and numerous destinations that the business travelers demand. The downside of this approach is the difficulty to find, or create, characteristically distinct customer classes in all (or most) markets in which a company operates.

Rigorous dichotomies require creative efforts in product design and/or distribution, which explains why development of a pricing strategy should begin early in the product development process. The market for photocopiers offers a good example. Back in the late 1980s/early 1990s, Xerox Corporation developed an interesting approach to

- monitoring, and
- measuring intensity and type of use of its machines

Having done that, it looked for market segments associated to each *use type*. The outcome enabled the company to develop a pricing schedule that led to a popular copier. With *use type* in mind, Xerox put into effect a product pricing structure difficult to resist.

For professionals the price structure included a usage charge for the number of copies made, in addition to a monthly fixed charge to rent the hardware. For the casual user, a smaller, less sophisticated but modern machine was designed and sold with bundled price.

Through this approach, customers who used the photocopies more intensively paid more for its services, but the increase was moderated. The charge also differed by type of copying. The more copies were made of an original, the lower the charge for a copy. Such pricing strategy requires a creative insight based on differences in buyers—not differences in technologies or production costs.

Other pricing algorithms are based on technological solutions. In early May 2010, Cisco announced a new technology to link up separate data centers in an intercloud or a "cloud of clouds" mode—just like the Internet is a network of networks. To the vendor's opinion, this justifies prime pricing because of *cloud broking**—which is a new and tough business.

Cisco, argues that there are barriers that could prevent computing from becoming freely tradable. Virtual machines† may travel easily, but the

* Meaning that customers switch between clouds or use several different ones.

† D. N. Chorafas, *Cloud Computing Strategies*, Auerbach/CRC, New York, 2010.

related data is much harder to move. Other cloud providers, however, have different opinions about pricing. In December 2009 Amazon.com introduced a new pricing option.

- Customers bid for its large unused computing capacity, and
- they get to run their virtual machines as long as their bid exceeds the minimum price needed to balance supply and demand.

Critics say that this solution, and its favorable pricing, does not lend itself to all applications. Virtual machines may be shut down at any time, when the spot price rises above the user's bid. Even so, it has made many user organizations think about computing in more economic terms, by asking what a given job is worth to them.

Another much more common way to segment a market for pricing is to require that buyers purchase a consumable good used in conjunction with a durable good. Creative pricing specialists have discovered numerous ways to do so. Kodak traditionally designed cameras to take only Kodak film. IBM leased its early computing machines requiring that they be used *only* with IBM cards.

The firms in these cases charged a very low explicit price for the durable goods. The price of supplies, however, carried a substantial margin, resulting in higher prices than those charged by competing sellers. Few customers are known to have discovered that the true price of the durable goods was not the low explicit price, but the low explicit price plus the extra cost of supplies.

Options is another pricing strategy. Manufacturers of motor vehicles are constantly developing new products and options. The prerequisite is to properly identify buyer groups and what they are willing to pay for specific benefits. By deliberately designing benefits that buyers find attractive auto companies can price them to earn more on each model run.

In conclusion, product pricing should not be monolithic and reverse engineering of costs is a good way to go about it. A valid solution requires that customer differences are identified early in the product development process. In mechanical, electrical, and electronics products one of the crucial differences is sensitivity to maintenance and repair services—which for some customers is integral to the product and they want it contractually tied. Such bundling is one of the criteria that can be used to segment markets for pricing.

## 12.5 PRODUCT PRICING IS NOT A SCIENTIFIC DISCIPLINE

"The reason why corporations get big," said the chief executive of a major industrial concern in a meeting, "is because people want to buy their products. The reasons people want to buy their products is that they are good and they are priced right." If this is the case, then a question automatically comes up, "How do we price our products?"

Section 12.4 brought the reader's attention to specific cases. What we now need to do is to examine the more general case, which, as people with pricing experience admit, is generally easier to answer in a time of general price stability—though even then measuring actual prices, and deciding on new product prices, poses extremely challenging financial, commercial, and technical problems. To mention only three:

- Existing price indexes do not take into account quality improvements and do not adequately measure discounts.
- The fact that moves of the competition are not always rational impacts on product pricing in an important way.
- In principle, though not necessarily in fact, the price of a product or service should correspond to the benefit it brings to its user.

The third bullet looks too idealistic, but there are real-life cases to back it up. In 1869, George Westinghouse, an American inventor, turned the pneumatic notion that was employed in drilling the Mont Cenis Tunnel into a brake for use on trains. The Westinghouse concept of airbrakes was simple and not at all costly—yet of great utility. In pipes running under the train, compressed air held back pistons.

- In the event of a release of air pressure, the pistons slammed forward.
- The effect of this action was to drive brake pistons against wheels.

With pneumatic brakes, a 34 m train going at 50 km per hour could be stopped in 170 m, a feat never before attained. This encouraged the idea of scheduling more trains, running them more closely spaced than had previously been wise. In turn it required better signaling solutions.

- Better scheduling made feasible significant benefits from trains and lines

- A new market opening up, like signaling brought more advantages that were not reflected in the airbrakes' price

As this and plenty of other examples demonstrate, there is nothing really "scientific" about setting prices. The different theories about monopoly, oligopoly, and free competition might be good as general frames of reference, but they are fairly void of substance when it comes to fixing price levels.

Moreover, pricing solutions inevitably involve errors in economic calculation not so much because of the deviation between actual and correct prices (whatever this means) but because there are not (and never were) any "correct" prices with which market prices perversely refuse to coincide. Prices are set by market leaders, the way it suits them best.

Every industry and every product has a market leader who sets the price. What the other firms can do, if they wish to survive, is to "follow the leader" and adjust their internal costs in ways that allow them to make profits. I had a professor at UCLA who had been senior executive at Ford Motor Company. He visited Ford frequently. In a graduate seminar, a colleague asked him, "How does Ford set prices?" He answered, "The auto prices are set by General Motors. Ford reads them and tries to do better than that!"

This was Honeywell's policy in the early 1960s with computer pricing set at 10 percent below IBM's price list, with another 10 percent offered in more power than IBM's price for corresponding equipment. Along with that came a "Liberator" routine to assure portability, and a determined effort to show the prospect *his* interest in going Honeywell's way.

The reader should as well notice that in pricing its products and services a company does not only confront the decisions and actions of its competitors. It also has internal problems to sort out, as its new products compete with its own older product because their functionality overlaps. There are as well many cases of products brought under the fold in sequence to mergers and acquisitions, which are poorly integrated with the acquiring firm's product line—while management might have failed to rationalize the company's own products in the first place.

A classical example has been the integration of car models by Alfred Sloan in General Motors, in the 1920s. The different brands that came under the GM umbrella were competing with one another in market appeal, targeted market segment, and price. Sloan reorganized them with Chevrolet at the bottom and Cadillac at the top—with Buick, Oldsmobil, Pontiac in between.

In the 1960s as consultant to Olivetti, I was confronted with a similar problem, and Sloan's masterly solution was the paradigm to follow. In the late 1980s this problem reappeared as the company's three product lines: word processors, accounting machines, and small computers merged because of technological evolution.

- This gave a disorderly product line, with duplications and triplications as shown in the top half of Figure 12.2.
- The overlapping products were screened, revamped, and restructured so that only the smallest possible overlap—necessary for commercial reasons—has been maintained. Retained products were organized in five product lines, L1 to L5.

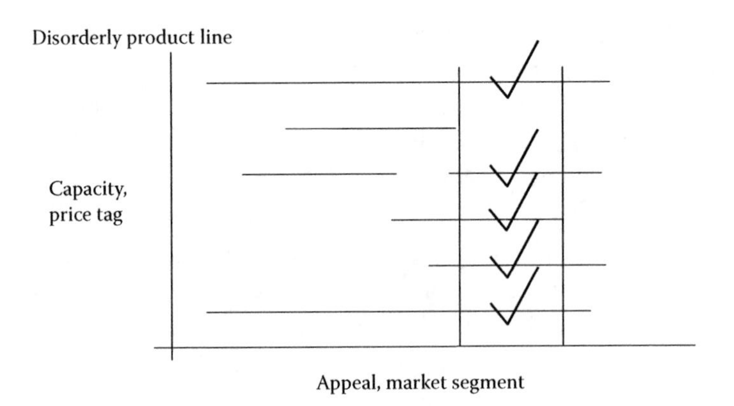


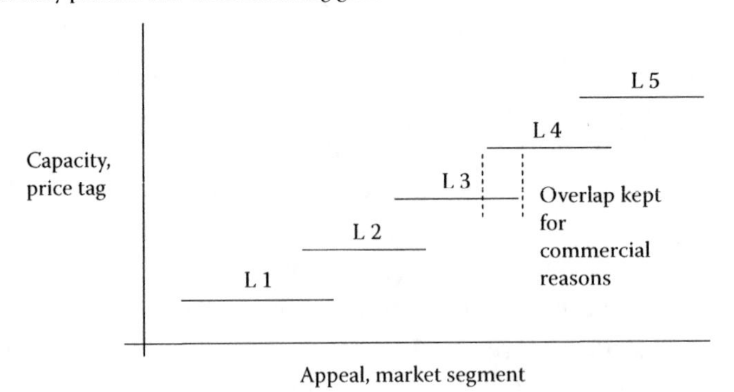


**FIGURE 12.2**
A company's product line must be orderly with well-chosen market objectives and corresponding price tags.

Plenty of reasons influencing pricing decisions, and the pricing mechanisms itself, pose difficult problems of adjustment for all sorts of producers. In times of inflation, cost increases can be accommodated through higher prices. With globalization, higher prices can lead a company out of the market. At the same time, it is obvious that unless a manufacturer can sell his products for more than it costs to produce them, he will soon cease to be in business.

Each product pricing case has to be studied on its own. In a period of low-volume sales, a manufacturer who attempts to recover his entire R&D investment and overhead costs out of the unit selling price will price himself out of the market. If he distributes his overhead on the basis of a temporarily high level of demand, he will mislead himself into believing that he will soon be out of the tunnel.

Cost accounting is providing due vigilance, but not every manager appreciates it. In the 1920s, anxious to keep sales expanding Henry Ford had a habit of lowering prices without consulting his cost experts. In the background of his policy was the optimistic belief that, if challenged, his engineers could discover new methods to lower costs.

That policy worked remarkably well for many years. When this happens, it creates the innovative climate in which ideas best prosper. We cannot order someone to be creative, but we could follow a policy that encourages creativity. The question is: For how long will such a policy really work? In principle, creativity can go a long way toward cutting unnecessary fat, but cost cutting alone cannot work miracles. Several issues come into play:

1. The impact, extent, and pace (and, therefore, investments) in R&D
2. Steady modernization, automation, and improvements in manufacturing efficiency—hence, investment decisions
3. The cost of commercialization, from the construction of a network of outlets to the training of salesmen, sales support, and after-sales services
4. A policy of very low overhead, which is not in the culture of most organizations

Product pricing is not a scientific discipline, but there exist management principles that help in doing it more rationally. The British say, "It is famously hard to govern," and governing prices has been one of the tougher jobs. However, creative minds, which as Winston Churchill once remarked "see opportunities in every difficulty, rather than difficulty in every opportunity" are able to find commendable solutions.

## 12.6 THE NEED FOR FORMAL PROFIT PLANNING

Well-managed companies are keen to put in place and maintain a product pricing mechanism characterized by a sound methodology and possessing continuity of action. It must be able to not only capture events relating to market forces and their pricing aftereffects, but also project well in advance the likelihood and severity of their occurrence. A multistage approach can help:

1. Start with market targets and project a price range from the point of view of direct competition.
2. Examine the dual aspects of product image and company image, and evaluate the impact of pricing policies on the longer term.
3. Establish the guidelines of a pricing policy, always keeping in mind the pricing policy of competitors.
4. Study the marketing capabilities for launching new product(s) and compute the type and number of skills to reach desired results.
5. Continue to work on a pricing strategy on the basis of squeezing internal costs and watching the competition's challenges.

Two or three decades ago, the approach many firms took in exploiting the synergy between product prices and profits was rather casual. As long as they were getting a good cash flow they did not worry too much about the fact that pricing policies are indivisible from the company's market image—and this promotes the need of formal profit planning.

Still this is not universal practice. When it comes to planning for profits, many industrial firms and financial institutions remain unconvinced. They will agree readily that profit planning is an exciting concept; they concede its logic; but in practice, they argue, it just doesn't seem to work, "at least not in our business." What explains this? Observation suggests that the answer lies on four misconceptions that handicap profit planning efforts:

- Profit planning is unrealistic.
- It is a job for staff specialists.

- It does not concern down-the-line managers.
- It cannot really be applied in all its aspects.

Each of these misconceptions might contain a grain of truth, but if adopted any one of them can paralyze the profit planning effort. While all future conditions cannot be predicted with the same degree of accuracy, companies find in these misconceptions the perfect excuse for mediocre performance. Merely aiming for "increased profits" is not enough, and no board of directors should accept this as planning objective. *If* we aim at increased profits, *then* we should ask ourselves the following questions:

- "Increased" in terms of what measures? By what standards?
- How much improvement and by which point in time?
- What are the assumptions regarding the reason why our wish about increased profits will be met?

These questions may sound obvious, but it is surprising how often they go unanswered because of the lack of a formal profit planning discipline. Also, since profit planning integrates the pricing of products and services, the groundwork involves determining market trends, estimating market demand, examining the competition, and providing ranges of variation.

A methodological approach to profit planning is a philosophy, attitude, and practice that must be conducted at all levels of the organization, with management providing the framework within which it can be undertaken most effectively. In addition, price studies and the profit planning efforts may be pointless unless line managers are held responsible for profit results, which in turn means that they must play a key role in profit planning. Line managers must not only be involved in the plan's development, but also support the assumptions that underlie it.

- They must view it as *their* plan, and
- as an effort aimed at achieving *their* objectives.

This does not change the fact that at the end of the day responsibility for effective profit planning lies with top management that sets corporate goals. What it means is that managers down the line, with their detailed knowledge of operating conditions, are best qualified to translate these broad objectives into specific planning programs. Only if they participate in setting profit goals will they be truly committed to achieving them.

The line of business a company is in evidently impacts on its profit planning methodology. Other issues, too, matter. A distinction must be made between equipment for sale and for rental. Equipment for rental, a policy which has been widely followed with mainframes, has financial carrying costs. Therefore, IBM has been extremely careful with what in company jargon is known as *capitalization.*

Capitalization = direct labor and direct material

Capitalization has been classically set at IBM at a small fraction of the list price—which fraction engineering and production must meet. A rented unit at client premises carries its own capitalization. At the same time, however, it guarantees a steady cash flow as well as the manufacturer's presence on the client's premises. It may also be that "rental" helps the manufacturer save taxes if he can ensure a 10-year minimum life for his equipment.

Whether selling or renting his equipment, a manufacturer must recover his costs and make a profit. Cost-oriented practices have found widespread use in manufacturing and marketing; cost-plus pricing practices are an example. They have been, however, criticized by economists on several grounds.

Under conditions of competition, economic theory predicts that unit prices will tend—in the long run—to be equal to total average unit costs, including a normal profit, though no necessary relationship exists between price and costs. Associated with this is the question of accounting practices, the argument being that (particularly in a time of inflation) accounting costs utilized in cost-plus pricing are both theoretically and factually unsound:

- They do not allow for many implicit or opportunistic costs.
- They do not properly handle items such as depreciation reserves, dividends on preferred stock, and capital gains and losses.

Furthermore, the reaction of competitors is very often a crucial consideration imposing practical limitations on pricing. Winning our competitor's customers with salesmanship or superior products is one thing; attempting to win its customers with price cuts and through price wars (Chapter 13), is quite another.

Last, but not least, more than anything else, pricing is a matter of the bosses' decision. As Ralph J. Cordiner, a former chief executive of GE once stated, "The timid, half-hearted commitment will never become a significant source of profit." A decisive commitment can come only after the leader of the company shows his skills, and this classically requires a good deal of experience in the market. Success stories always reflect good management at every level.

# 13

# Computer Price Wars

## 13.1 AN INFLECTION POINT IN THE COMPUTER INDUSTRY

The statement has been made in Chapter 12 that there is nothing scientific in product pricing. In years of "business as usual" companies try to rationalize the price list of their products, and at the same time capitalize on market inefficiencies. But major market changes, often induced by technology, have a snowball effect leading to an inflection point where price wars become all inevitable.

With computers, this happened in 1977, a change in industry pricing practices promoted by the plug-compatible manufacturers. The saga started with *Project M* by Japan's Hitachi and Fujitsu who were at the time the country's foremost companies in industrial electronics,* and whose strategy was to eat up for breakfast IBM's market.

The all-powerful Japanese Ministry of Foreign Trade (MITI) that targeted exports and directed with an iron hand the Japanese companies' industrial policy, financed a project to emulate IBM's mainframes—with a view to conquer a big chunk of the global computer market. In the United States, the partners of Project M have been Amdahl† and National Advanced Systems (NAS), but soon developed a whole industry behind this emulation of IBM mainframes, the so-called *plug-compatible manufacturers* (PCMs) and their machines that run on IBM operating instructions.

Amdahl was joined by NAS and Magnuson, which were the first plug-compatible names, but soon there has been a golden horde of PCMs. They were run by young people with a goal, and they managed to change the

* Fujitsu was set up by Siemens when in the mid-1920s the Japanese government asked the German for assistance in developing an industrial electronics industry.

† Of Sunnydale, California, with backing from Fujitsu.

style, if not the structure, of the computer industry. Slowly but surely the plug-compatibles established themselves as a permanent force.

This did not happen overnight, neither did the PCMs cut as sharply as originally thought into IBM's hold on the mainframe market. Rather, they capitalized on a market-share shift. Instead Sperry Univac, Burroughs, NCR, Honeywell, and other IBM competitors* saw their share of computer systems in the 1000 largest American computer installations shrink.

What really allowed Amdahl and the PCMs to gain a foothold in the U.S. mainframe market were the inevitable weak spots IBM and the BUNCH had left in their poorly organized product line. Big Blue had planned to introduce its ambitious Future System in 1975, and for once it was a design incompatible with its previous computers. In the end,

- an internal struggle was won by IBM marketeers who decided this change would be too much of a wrench for IBM users, and
- the Future System was junked, which meant that IBM had to extend the life of its old products by announcing the interim 3000 series computers

This 3000 series was a relabeled version of the 370 series, which itself was an upgraded version of the 360 that dated back to the mid-1960s. All of them were hard core mainframes with zero-point-zero innovation—while not only the PCMs were out of blood but also minicomputer makers like DEC were encroaching into IBM's customer base.

The PCMs were ready to step in and Big Blue's fallback to a defensive position could make no use of the old 1910s strategy of the "broken down cash register"† IBM had revived in the mid-1960s to confront the onslaught in scientific computing by Control Data. Something new and more powerful had to be invented to save the day. In March 1977,

- IBM disclosed price cuts of up to 30 percent on most of its 3000 computers, and
- IBM introduced a new central processor, the 3033, to compete with the top-of-the-line offerings by Amdahl, a move induced by stock market tremors ripping through its stock price

* Nicknamed at the time as the BUNCH, from an anagram of their labels.

† An NCR strategy engineered by Thomas Watson Sr., its marketing director, to take over the cash register market from Standard and other established manufacturers.

One voice in the stock market crowd declared that by being more confident about its antitrust position after its victory in the Cal-Comp case, IBM appeared to have decided to slug it out rather than see its market share dwindle. The PCMs strangled as the price cuts unsettled their strategy to match the lower level IBM prices, and do better than that. Prices plunged both for computers and computer stocks.

Amdahl countered with 29 percent markdowns of its own, and promised two new computer processors, one of them 1.5 times more powerful than IBM's 3033. But equities of the entire mainframe group—Burroughs, Honeywell, Sperry Univac, as well as Itel and Amdahl—quickly sold off as investors tried to reconcile existing profit margins with 30 percent lower prices. Ironically, foremost among the financial casualties was IBM itself, which had paid $280 per share for 2.5 million shares of its own stock just weeks before the market buckled under the weight of its own pricing announcements.

Price cuts' aftereffects disproved Big Blue's widely heralded forecasts. Yet, 1977 was to be a great year for computers in orders and installations, none other than IBM chairman Frank T. Cary had told rows and rows of shareholders in the New York Coliseum in February of that year. Cary had also emphasized that rentals and sales were off to as good a start "as we have had in recent years." (One of those attending, a lady all in black, revealed that she had dressed to mourn IBM common, then 30 points below its 1977 high.)

The PCMs were in no way better off—even if a spokesman for Itel crowed that, "It hasn't affected our firm orders, not at all." A spokesman of Amdahl was no less misleading. "We haven't cut back one iota on any item of expense since the announcements. It's a full steam ahead." As for IBM, Frank Cary had said that he expected these lower prices and the accompanying increased performance will expand the market—and the company's profits. Less means more, in New Talk.

Realists saw it under a completely different light. Excess computer capacity in users' hands stood at the lowest level in years, and that was good for sales. But deep price cuts depressed revenues. A spate of new-product announcements were a bullish omen, some observers said, as buyers were likely to do their shopping early in the product cycle—but huge price changes brought uncertainty to the market, and the failure of software to keep abreast of hardware created its own sales inertia.

## 13.2 PRICE WARS AND THE STOCK MARKET

As it should have been expected, the price war, which in early 1977 raged in the computer industry was wreaking havoc on the stocks of the big computer manufacturers. In the first 2 weeks after IBM's 30-percent price cuts, computer stocks present the pattern shown in Table 13.1.

Burroughs, which cut its prices to match IBM's fell by nearly 10 percent, while Honeywell and Sperry Rand that were expected to chop prices shortly to remain competitive, were down by 8 and 9½ percent, respectively. True, the overall stock market had taken its own lumps during the same period. "But the computer stocks," noted John J. McManus, vice president of Kuhn Loeb & Co., "have acted way worse than the market place."

Wall Street's immediate worry was what the computer price war will do to industry profits, as financial analysts cut their projections of 1977 pretax profit margins by about half of one percent for each of the big computer companies. In addition, some knocked at least 60 percent off their 1977 estimate of IBM's net earnings, and something comparable off their estimates for Sperry Rand and Burroughs, who were then Big Blue's two main competitors.

The following reactions were registered in terms of selling prices of computer stocks. In a dynamic market like the New York Stock Exchange, these

**TABLE 13.1**

Equity Prices Plunge after IBM's Announcement

| | Stock Prices | | |
|---|---|---|---|
| **Company** | **March 22, 1977** | **April 5, 1977** | **Percent Change (%)** |
| Amdahl | 27 1/2 | 25 3/4 | −6.36 |
| Burroughs | 68 1/8 | 61 1/8 | −9.91 |
| Honeywell | 51 1/4 | 47 1/8 | −8.05 |
| IBM | 285 1/2 | 274 1/2 | −3.85 |
| Itel* | 14 7/8 | 13 3/4 | −7.56 |
| Sperry Rand | 39 1/4 | 35 1/2 | −9.55 |
| Dow Jones Industrial Average | 959.96 | 916.14 | −3.66 |

* Contrary to the other companies in this list Itel did no manufacturing. National Semiconductor, was making two models for Itel's exclusive distribution, the AS (for "Advanced System") -4 and AS-5, aimed respectively at the IBM-148 and -158. By selling its wares for cash, Itel had shifted the risk of technological obsolescence from itself to its customers.

reactions reflected the concern and appreciation by analysts and investors on what is (and may be) going on. But there was no consensus. To provide the basis for further thought and discussion, I present three cases that contradict one another: Bear, Bull, and Median.

- *Bear case*: Technology has run amok. Cheap semiconductors available to all mean new competitors, more price cutting, declining profitability, greater PCM competition, and IBM price cuts were the evidence, and they were only the first step.

Analysts espousing the bear case argued that the market for mainframe is saturated, while capacity is being delivered faster than users can absorb it. Plug-compatibles and minicomputers are taking market share away from Big Blue, and the Japanese will eventually enter the U.S. market aggressively (it never happened).

- *Bull case*: Declining prices and increasing performance always has and will continue to stimulate market expansion. Therefore, this is a normal and necessary part of the business. The current concern is due to "tail end" of product cycle effects.

Bull case analysts had some advice for computer vendors. They said that key to maintaining pricing and profitability will be product differentiation incorporated into next generation computer equipment. The market remains elastic to price. There are many new applications. Technology gains are key to growth, and they are expected to continue. PCM competition is temporal, opening new markets.

- *Median case*: Hardware is no longer a barrier to entry into the information system market. IBM still controls software and system architecture, and can therefore provide for product differentiation. The question is to establish whether concern will ease or the situation will become more complex with a continuing price war.

Computer vendors had a different set of concerns. For most of them the worrying problem was that because of technology and competition, prices were declining too rapidly and they had to cut down costs because declining costs were the key to an expanding market and to holding their integral share.

Both among vendors and among analysts there have been as well three views of profit margins: The *bear case* stated that increasing competition using cheap semiconductor technology causes price wars, which destroy industry profitability. The thesis of the *bull case* was that computer makers can continue to maintain profit margins via product differentiation and, most importantly, account control.

The *median case* stated that profit margins will be under pressure due to increased hiring and inflationary impact of employees on benefits, as computer vendors fought to hire and retain the best talent. (This was particularly visible in Europe.) Pressure could also be expected to come from the impact of near-term IBM tactical moves relative to the product cycle.

A growing group of market analysts were of the opinion that it was not only the price cutting that routed computer shares. Institutions were rotating their investments selling off lower-yielding growth stocks in favor of stocks offering a high total return. In fact, to spur that interest, IBM had both raised its dividend and offered to bolster its market price by buying 2.5 million of its own shares from investors at $280 a share. Many questions, though, were still looking for an answer:

- Who *really* started the recent price war?
- How will it affect the customer base?
- What will it cost to the computer industry?
- Where will it end, and by then who will survive?

The answer to the first question was not that easy, as IBM claimed it did not fire the first shot. True, Big Blue had long been expected to mount a counteroffensive against Amdahl and the Itel-National Semiconductor alliance, which had enjoyed surprising success selling IBM-compatible computers.

The inroads were particularly pronounced with machines that used IBM programs in the middle and top range.* But several financial analysts claimed that the latest round of price cuts were part of the grander scheme of *creative destruction*. Costs had been falling and new technologies were born. Using large-scale integration, Amdahl, with assets of $84 million, had succeeded in manufacturing and selling a computer rated superior in price and performance to the IBM 168.

* Since the early 1970s top IBMers tried to guess who would make the first plug-compatible entry. They guessed wrongly on Texas Instruments and the Japanese landing in the American computer market with their brands.

According to then Burroughs Chairman Ray Macdonald, since late 1975 a general 2-for-1 boost in mainframe price-performance ratios had become standard. "I'm not saying Wall Street is right or wrong," Macdonald commented in a press interview as Burroughs was making another new low in New York. "I'm just saying that it's two years late in recognizing present trends, and perhaps overreacting."

"IBM's moves," insisted Burroughs' Macdonald, "will have no impact on our ability to compete nor will they have a negative impact on our anticipated profit margins." Burroughs stated that it continued to regard $5.30–$5.50 a share as reasonable in 1977. And at Sperry's Univac division bookings were up 36 percent in the first quarter of 1977; but this was before the start of the computers' price wars.

All told, a clear-eyed observer would have noted that whatever the future held, the belligerents seemed considerably less worried than the bystanders. Control Data, which was planning for some time a limited foray into the 138/148 (medium range) plug-compatible market, said it was getting ready to unveil its product, evidently undeterred by IBM's latest move. I haven't seen any margins depressed, observed Clarence Spangle, head of Honeywell's computer division. In fact, we've seen ours improving.

## 13.3 TOUGH COST CONTROL IS THE BEST WAY TO FIELD OFF COMPETITORS

Discounted for bravado, the sentiments quoted in Section 13.2 largely rest on two unarguable facts: First, that manufacturing costs had generally plunged faster than prices. Second, that lower prices called for more customers, higher revenues, and better earnings. That was good news but it affected the computer vendors in an uneven way.

Analysts who shared that line of thinking pointed out that manufacturers' costs of the typical intermediate to large processing system were projected to fall by 80 percent in the decade of the 1970s. Not only were components' costs dropping, but also, and more importantly, the comparative cheapness of the computer gear altered the historical design ground rules—making it technically and economically feasible to rethink how data systems should be projected. (See also the discussion on cost control in Chapter 15 and on overhead costs in Chapter 16.)

There was another way of looking at this situation. *If* fast advances in technology were brought to the foreground through an aggressive policy of acquisitions, *then* the results could be visible in dollars and cents. IBM had at the time $6 billion in its coffers and was looking for profitable ways to invest them. Instead, it chose the price war.

- Viewed under the perspective of dollars and cents, the question "Can IBM endure it?" could only be answered in one way. "If IBM can't, who else?"

In terms of technology, Amdahl made a powerful 470 V/6 computer that operated on IBM's software and could easily replace large and more costly IBM units. But the installed base was too small. Amdahl had placed about 50 of its computers, and managed to grab over 20 percent of new large-scale system shipments; but the size of IBM's client base was very much larger, and Big Blue derived 80 percent of its annual sales from it.

Seen from an engineering viewpoint, Amdahl's line of thinking was correct, but in 1977 the pundits goofed in the specifics. Their hypothesis had been that (unspecified) electronic elements can be substituted for many of the functions previously performed by other, more expensive methods—specifically software painstakingly written by people. The decades that followed saw to it that this became an unsubstantiated projection.

In that regard, some computer company executives were much more realistic. "One of these days, before the end of the next decade," said at the time John V. Titsworth, executive vice president of Control Data, "hardware's all going to look alike and be inexpensive. What we'll all be selling then is applications and system operating software, plus expertise in particular application areas."

In other words, Control Data correctly projected that technology was due to become more homogeneous and a "commodity." Within 13 years (1990), Titsworth guessed, with IBM and its competitors hard-pressed to distinguish their wares. This, too, had a role to play in cutting down production costs—while at the same time Japan cast a shadow over the more or less distant future.

Indeed, some analysts suspected that IBM's action in March 1977 was aimed less at Amdahl than at Fujitsu, the California concern's largest backer and a major producer of computer systems, telecommunications gear, and electronic components. (Fujitsu's computer sales in the fiscal year that ended in March totaled $691 million, a sum placing it ahead of NCR and Control

Data, but well behind Digital Equipment Corp., Sperry Univac, Burroughs, Honeywell, and of course, IBM. But in their home islands, where Japanese manufacturers account for two-thirds of computers installed, Fujitsu led IBM in market share, roughly a 21 to 27 percent.)*

In the American computer market itself, the PCMs were showing the sceptics that they can not only keep up the price pressure, they could also forge ahead of IBM technologically. The confidence that Gene Amdahl and other executives of PCMs displayed was probably rooted in their intimate knowledge of IBM's technology and marketing practices as the PCM industry was full of IBM veterans. In terms of manufacturing, the credo was that *if* growth and falling costs are the computer industry's tradition, *then* that industry must get organized not only to survive but also to thrive in that new environment.

This reinforced the need to be in charge of costs. The better way to look into the control over manufacturing expenditures is in the light of results obtained by the start-up PCMs, which is to examine not only their innovative features but their strategy, way to marketing and sales, where the formerly unreasonably fat margins were most significantly reduced. The PCMs attacked the old, established computer companies in three fronts:

1. Greater efficiency
2. Better market appeal
3. Capitalization (DL+DM) costs†

With the plug-compatibles and the fast growing minicomputers industry, product life cycles that were classically 10 to 15 years, suddenly became five years. "If we don't obsolete out products, somebody else will," said the CEO of a company to whom I was at the time consultant. In the manufacturing floor itself robotics and other technological processes had changed the picture of operations the way we knew them. This refocused capital investment planning to meet short-, medium-, and longer-term objectives.

- A rational capital investment planning requires a comparative evaluation of alternative projects that use company capital.

---

* Japan was the only country where U.S. computer manufacturers did not hold a commanding lead.

† See Chapter 12.

- A specific duty was to assure that reliable and systematic methods are used to describe and evaluate various investment proposals; a significant change in the production floor.

In terms of its effect on pricing, a renovated manufacturing technology worked both ways. A conversion to robotics calls for significant capital outlays, but the manufacturing automation it promotes helps to drastically cut personnel costs. Still plenty of homework was needed to assure investments would be financially justifiable, including integrated evaluation procedures.

At least in some of the computer companies, budgeting and accounting were once two completely separate functions, the former used for planning and the latter for assessment. Being separate processes, there was little possibility of relating the results of one period of budgeting plans for the next. Competition saw to it, however, that in the 1970s and 1980s they became integrated into one coordinated system:

- This allowed for a common basis for data gathering and evaluation.
- More rapid assessment of short-term data occurrences helped in the adjustment of longer-term projections.

The better managed computer companies have also been keen to assure that directives related to cost/effectiveness are carried out to establish a link between cost control methods and the behavior of individual cost centers. This need was particularly appreciated by managers who knew that the control of costs requires a timely and effective comparison between costs determined by activity, and costs determined by capacity, particularly idle capacity

Under an integrative planning and control perspective, financial plans epitomize the idea of cost-versus-actual evaluations, providing the standard against which performance is ahead of the curve, and making feasible corrective action before results get out-of-hand. Two different computer companies to which I was consultant to senior management applied this principle.

Precisely the same principles applied to distribution costs. They, too, had to be cut with a sharp knife if a computer company, which by necessity engaged in the price war was not going to get busted. The real cost of distribution includes much more than what most companies consider when they attempt to deal with budgeting. A major decision regarding the distribution network affects in a significant way the cost of doing business, and also relates to other cost issues. Marketing and sales proper aside, a number of elements may prove critical in evaluating the impact of distribution approaches on total costs of sales. They include warehousing, inventories, supply chain, channels of distribution, and maintenance.

## 13.4 WAREHOUSING, INVENTORIES, SUPPLY CHAIN, CHANNELS OF DISTRIBUTION, AND MAINTENANCE

Warehousing is required to provide a reliable storage service through the company's chosen channels of distribution. The facility can range from one in-plant warehouse to a multinational network. Quality of service usually becomes better as the number of warehouses is increased (at least up to a point) but costs might go out-of-hand as the number of units multiplies, if management does not watch out.

Several structural aspects must be considered. Other things equal, as the number of warehouses increases their average size tends to decrease, and so do the number and quantity of items. In cases, this may reduce the efficiency of service to customers. Change in the three variables—number, type, and location—of warehouses affects both customer service and costs.

A major issue connected to transportation costs is the *means* being used: train, truck, ship, and air. Changing the number or location of warehouses alters the picture of the transportation requirements and associated costs. An increase in the number of warehouses may initially reduce total transportation costs, but past some determinable point, the cost trend may be reversed because of the decreasing ratio of carload to less-than-carload tonnage.

Large inventories cost money in terms of invested sums, insurance, possible taxes, real estate, pilferage, and custodial services. Depending on the business, inventory costs may range from 10 to 30 percent of average annual inventory value. Customer services can be improved by keeping a larger inventory, but this also increases the cost for carrying that inventory and it is reflected in prices. (Dell made a fortune with personal computers by cutting to the bone inventory costs.)

Optimization rather than sharp inventory reduction is the best policy, because inventory carrying costs are closely linked to warehousing costs and customer service (see also Chapter 14 on inventory management). The best way to approach this subject is to control inventory levels dynamically through a steady simulation of sales, service, and production capacity. On the other hand, the more complex the distribution system, the slower is the inventory turnover. An associated risk is inventory obsolescence.

The term *supply chain alternatives* refers to both business partners and the company's own manufacturing networks. Production costs vary among plants and as a function of the volume produced at each plant. The decision of which market should serve which customers must give weight not only to transportation and warehousing costs but also to quality considerations, time delays, and labor costs. The specific solution will vary with the type and size of the business; nevertheless, an effective supply control system needs the following:

- A good plan, with standards able to guarantee satisfactory performance
- A system for evaluating actual performance periodically against the plan, including both a clear explanation of why variances have occurred and a forecast of future performance
- An "early warning" system to notify management in the event that conditions warrant attention between reporting periods

Channels of distribution choices profoundly affect the nature and costs of a company's sales organization, its selling price and gross margin structure, and the commitments to physical distribution facilities. In turn, these affect production and supply. Communications costs vary with the complexity of the distribution system and with the level of service provided, including items such as order processing, inventory control, payables, receivables, and shipping documents. Such costs rise as more distribution points are added to the system.

Distribution decisions can affect costs otherwise incurred by suppliers or customers. When a retailer creates his own warehouse(s), this may free suppliers from packing and shipping small quantities or from maintaining small local warehouses in the field, but it is not necessarily certain that he will recoup these costs by negotiation with the suppliers, unless such negotiation has been done beforehand and has become one of the conditions for the network.

Field maintenance is a whole issue with great impact on the computer manufacturer as vendor of high quality products. Technology-intense gear has a voracious appetite for after-sales service. Typically, a field maintenance organization supported by the vendor bears two overhead costs:

- The traditional burden of supervisors, managers, facilities, and so on
- The idle time needed to ensure a timely response to customer calls

Even if machines have been designed with long mean time to repair (MTTR) someone has to be available just in case. In the data processing industry, expectations for response time have been classically between 2 and 4 hours—a rather heavy burden, since this means plenty of idle time for maintenance engineers. Traveling to and from the customer's premises is also time-consuming, hence, the interest in online diagnostics and self-maintenance, which means modularity, standard exchanges, and diagnostic ability through "do-it-yourself" kits with parts and instructions.

## 13.5 HALF-BAKED SOLUTIONS HAVE SHORT LEGS

IBM could afford the late March 1977 30-percent price cut, but this was not true of all of its competitors. Before too long, high-flying Itel was forced out of the computer-leasing business and went bankrupt a year later. Amdahl's profits went south, as customers began to lease rather than buy in anticipation of new, cheaper, big computers from IBM. And as we saw, the stock prices of Burroughs and Sperry Rand plunged.

At the same time, however, the price-cutting hurt IBM's then stellar profit figures—leading Amdahl president John C. Lewis to the statement that "most of the bullets IBM fires hit IBM, too." Lewis had reasons to be upbeat. He and his colleagues were proud of their accomplishments as its market success was, at that time, beyond doubt.

The company that reported its first full year's results in 1976, had shipped 42 of its 470 V/6 computers. In the three months that ended on April 1, 1977, Amdahl had earned $10.2 million—$5.4 million before tax-loss carryforwards—on revenues of $37 million. That was a pretty leap from the $1.4 million (half that before tax-loss carryforwards) racked up in January–March of 1976.

Ingenuity had flowered against formidable odds, and these results were reportedly causing a rash of dyspepsia in IBM's headquarters in Armonk.

"During the past several years," observed an Amdahl prospectus, which covered a 560,000-share sale of common equity, "there have been no successful entrants to the large-scale computer market, which is dominated by IBM, one of the largest and most powerful companies in the world. Precisely, since IBM had successfully checked CDC's progress with the 'Fighting Machines' stratagem. In addition, several major companies that offered a range of generally lower performance computers have withdrawn from the market."

Unlike RCA, to name a prominent victim, Amdahl built its computers to be perfectly, not nearly, compatible with IBM software. Its management did not believe in half-baked solutions. Moreover, unlike RCA, Amdahl struck out on its own in design and technology, offering a more sophisticated (air-cooled, where the 168 was water-cooled), more compact, and more powerful machine than its rivals.

And its computers were cheaper in cost/effectiveness terms. For the first time in its history, IBM was looking rather uncertain about the future, and its moves were *déjà vu*. Financial analysts started to question whether IBM's 3033 was for real or still another reincarnation of the knocked down cash register. To many, the 3033 appeared to be the "bridge component" to the next IBM generation, to the system beyond the basic 370.

"Up to now," one of the computer experts said, "the central processing units has been essentially one unit. But the next generation will break down the nucleus into molecular or even minicomputer size."* From IBM's perspective, this supermodular architecture—which may permit the user organization to add little blocks to accommodate its own workload—could help boost its purchase to lease ratio. But it was an alien policy to Big Blue, and it looked like a strange strategy to a firm, which for two decades had bet on monolithic mainframes.

For a vendor following a modular architecture, such a strategy would have helped to limit the capacity of plug-compatible makers to compete, by making it harder for them to take its software and run it on their hardware. This was not IBM's case, which led many computer industry watchers to believe that the next generation will rely to a great extent on microcodes; in a sense proprietary programming built into the computer.

* *Fortune Magazine*, January 25, 1982.

*If so*, then NAS, Amdahl, and the Japanese could be strongly tested. But it did not happen that way.

In the United States, not every computer market expert agreed with the aforementioned thesis. Some said that compatibility cut both ways. A user who deserts IBM for Amdahl, for example, can just as easily go back. This is not unusual as computers have been an industry where entrepreneurs live by the sword and die by the sword.

- The possibility of two-way traffic across a compatibility gateway, apparently had even begun to dawn on IBM's top management.

Over a period of time as the demand for computing power exploded and the mainframer had not had the capacity to produce all the computers it could possibly sell, the PCMs had a ball. But their advance also brought home the message that Big Blue could not continue to allow the erosion of its customer base. It was urgent that lost clients were helped to make a U-turn.

- IBM knew that it could capitalize on a well-trained sales force, and its red herrings.

Time and again, Big Blue salesmen were accused of trying to get the data processing manager of a computer buyer who had changed vendors fired. They did so by suggesting to his boss that his choice of a plug-compatible proved how bad his judgment was. Another frequent stratagem was dubious statistics on PCMs' equipment reliability, coupled with hints that service on IBM peripheral equipment might be reduced if a PCM mainframe is bought.

*If* the IBM salesmen pitch did not provide the expected scare at customer side, *then* a top IBM executive, including presidents of subsidiaries, would meet with the client firm's CEO to suggest that a saving of, say, $1 million was not that significant for a $6-, $8- or $10-billion a year company. And it was not really worth the risk of endangering a long-standing relationship with Big Blue.

- High pressure sales, however, have limits. IBM's weakness was that its clients may not believe in its interim solution, and look at it *as if* it was brought up to try to hold its position by using scotch tape.

A number of analysts felt that Big Blue was simply big headed and emboldened by a recent (at the time) legal victory in the antitrust action brought by California Computer Products. A Drexel Burnham analyst suggested that with this win behind it, IBM may be pushing a Trojan Horse hoping that nobody will perceive its strategy. The market did.

Along with nearly everyone else, Amdahl expected that IBM's next series of large- scale computers will be ready for delivery in 1980 or 1981. If the event occurred sooner, the lifespan of Amdahl's 470 series, including its new 470 V/7 processor, could be tragically short. This was by no means an impossibility because IBM has at times reacted quite aggressively when confronted with new competition.

Itel, the diversified financial services company that in October 1976 had widened its hostilities against IBM, was less exposed with its AS-5 series. In 1976, some 40 percent of its net profit came from the sale of IBM-compatible disc and memory products, a line that had mushroomed from $9 million in 1972 to $78 million in 1976. In the same stretch, the pretax net surged, while company-wide earnings grew 25 percent annually.

Outwardly, the PCMs were looking comfortable. However, for those computer companies inclined to worry, IBM had given more than enough cause. Neither was the Justice Department's antitrust suit against Big Blue, already in its eighth year, much of a consolation because rumor had it that Justice was dragging its feet.

In some respect, but only in some respect, worries had been overdone. New products, whatever the short-term uncertainty they may bring, are instrumental in turning the tables on a well-established manufacturer, and IBM's competitors were betting on innovation, not just on price cuts, in order to survive. This of course did not mean that casualties could be found. All parties conceded that endurance was at a premium.

## 13.6 USERS DON'T ALWAYS APPRECIATE THAT THEY ARE GETTING SEMITECHNICAL PRODUCTS

In the late 1970s the story of the 1950s repeated itself once all over again. In a short period of time, information systems vendors had to become aware that senior changes were occurring; be able to translate the meaning of these changes into specific product opportunities suited to their company's interests and capabilities; make decisions involving new product development or acquisition programs; and beef up their marketing to bring new products successfully into the market. To succeed, this had to be done (1) without much delay, and (2) on the basis of a firm plan of action.

Changes of a technological character have a most dramatic impact on companies, although they subsequently exert no greater overall influence than do other changes of a social, economic, or political nature. The reason for this initial misalignment of cause and effect is that frequently technological change appears to affect corporate growth opportunities in two ways, which work in synergy:

1. Through a business influence outside the realm of the company's control

For instance, technical developments featured by competitor companies, or the industry as a whole, may render a firm's products obsolete.* Alternatively, technical developments by competitors may open up new markets by providing novel uses. Generally speaking, newcomers have a free hand in the upside.

The principle is that technology breeds on technology. As new products continue being developed in technical laboratories, the scope of their markets broadens rapidly, and this in turn requires plenty of both engineering and marketing care.

2. Technological change created by the company by means of research and development is a "must" but it is not sufficient. Profits are generated by a well-trained sales force.

Compared to its challengers, in the 1970s IBM had a well-trained and motivated sales force ready for action. This was its competitive advantage. Big Blue was never at the edge of technology, as for example Sperry

* A case in point is that of electrical accounting machines (EAM) in the 1950s and of mainframes in the late 1970s condemned to decline with the rise of Vaxes and (in the following decade) of client-servers.

Univac used to be. IBM was strong in marketing and Univac in technical innovation. In plain terms

- Univac's, CDC's, and Amdahl's strength was *technical* products, and
- by contrast, IBM's strength was *semitechnical*, marketing, and hard sales

In the broadest possible sense, industrial products can be arbitrarily classified as "technical," "semitechnical," and "nontechnical," depending on the extent to which they have undergone technical change and the level of R&D breakthroughs associated to them. Also their apparent rate of technological obsolescence, which sometimes is more fiction than fact. Different marketing strategies are used for each of these classes.

Examples of product groups classed as "technical" and appealing to large markets are electronics, chemicals, drugs, synthetics, aircraft, missiles, control instruments, automation, new metals, and new energy sources. Examples of product groups classed as "semitechnical" is computer gear that has gone out of fashion, but is still around in computing centers. One of the better examples has been the whole group known as electrical accounting machines (EAM)—ranging from printers to sorters, collators, and keypunch equipment.

EAMs are today deadly obsolete, but in the mid- to late 1970s, at the time of the computer price wars, they were not only around in large numbers but also constituted IBM's foothold in data processing centers. Computers were, then, still housed in glass houses for visitors to see and admire. They were attended by white-clad priests who handled magnetic tapes sometimes bicycling between tape drives (a case I had in the early 1970s with Commerzbank).

The Amdahls and other PCMs upset all that cozy EDP living and its status quo. They also gave the impulse to start clearing out IBM's strongholds massively equipped with semitechnical products. (By contrast, examples of products classed as "nontechnical," but featuring high market potential, are packaged foods, apparel, clay and brick artefacts, as well as plenty of home furnishings and sporting equipment.)

Good marketing, however, can cover technological obsolescence up to a point. Beyond that there is a positive correlation between the rate of growth of a company and the advanced technical orientation of its products. Over half of the total sales of a high-growth company are what we defined as "technical" products. By contrast, half of the total sales of a low-growth group are of "semitechnical" products.

- These opposing directions of product characteristics have an evident impact on corporate strategy.
- Two times in the second half of the twentieth century, IBM made an urgently needed switch—just in time.

The first was in the mid-1950s when, having rejected the purchase of Univac, it found itself at disadvantage with the EAMs. The day was saved by its marketing director who framed the policy that the company was ready to install computers—but would not do so till the customer was ready for them. This led to the institution of applied science (system engineering), one of the most positive developments in IT.

The second time was in the early 1990s when a new CEO, an outsider, turned IBM's highly bureaucratic organization upside down, and out. For the first time in Big Blue's history, he engaged in aggressive acquisitions of other companies (particularly in software); radically altered the product mix; and renewed rows of management. But he retained the company's culture of mainframes, and this may poison client installations and system solutions in a cloud computing era,* as we will see in Section 13.7.

This deep love for mainframes, and therefore for continuing to live in the past is difficult to explain. Product renewal, aggressive pricing, and diversification activity in industry seems inextricably linked with advancing technology. Companies in a given growth field might be likened to the line-up of race horses at the starting gate. All have the opportunity to win, but, of course, only one can be first. The winner will be determined primarily by two factors:

1. The resources, capabilities, and training inherent in each horse.
2. The degree of skill, planning, judgment, and energy applied by the jockey.

And just as the choice of product fields is important, so is the choice of races entered. A company can enter a field early in its cycle of expansion potential, or when its past may have shown growth but its future shows

* D. N. Chorafas, *Cloud Computing Strategies*, Auerbach/CRC, New York, 2010.

maturity or decline. Both options are open but the outcome is loaded to the side of decline by sticking to the past, as Section 13.7 suggests.

## 13.7 THE DARK AGE OF MAINFRAME MENTALITY ENTERS CLOUD COMPUTING

In 1953 IBM run scared after Thomas Watson Sr. famously rejected the offer to buy Univac, saying that computers were a university professor's toy and the market for them was not going to be more than 50 years. But one of the old man's assets was to recognize his failure and correct it quickly. Eventually IBM overtook Univac and left it in the dust.

In 1963 IBM run scared because of the entry into what was then called "Electronic Data Processing" (EDP) of companies with sales a multiple of its own, like GE. Or with shrewd enough management to challenge its own; RCA was an example. Acting from a largely defensive stand, Big Blue allocated over a period of about six years (but by far the largest part of money over the three years from 1963 to 1966), the then colossal budget of 5 billion dollars to develop, manufacture, and market the 360 Series. Much of it was risk capital:

- The cost of hardware R&D was roughly $500 million.
- 360 software proved the hard nut to crack, to the tune of $750 million.
- The lion's share of the balance went to the upgrading and rebuilding of the marketing and sales network.

Other investments went to manufacturing where the element of risk plays a rather minor role. Plenty of money was needed for the development of new factories to make the 360 series from LSI to final assembly. Risk capital was again in the picture, as the company financed a 40 percent increase in sales and sales support personnel within three years.

For all of the above reasons, in the early 1960s when IBM management decided to plunge in the $5 billion, this was considered to be a "gamble." Domestic and World Trade IBM business did a yearly turnover of about $2.5 billion—about half the money to be invested. Top management (read Thomas Watson Jr.) bet the company. But it paid off. The guess was that with this investment by the late 1970s, IBM will be doing an estimated

$15 billion turnover, and will lock and hold the No. 1 position in the global computers industry.

This is precisely what took place and it lasted for years till in the late 1980s/early 1990s IBM nose-dived (more on this later). Critics say that such a policy was not foresight but a mixture of arrogance and greed that brewed in the 1960s and 1970s, the years of fat profits. It was as well fed by the incredible incompetence characterizing the other established computer vendors.

The main IBM competitors who, prior to the PCMs and the minis were all mainframers, became culturally gridlocked, thick-headed, and increasingly lost in a fast-changing marketplace. The irony is that success spoiled IBM management and Big Blue, too, turned into a first-class example of gridlocking and thick-heading.

Friendly chief information offers at user organizations contributed to the hallucination. The mainframes were *über alles*, and customers were offered plenty of choices—provided they were mainframes. As time went on, convinced of the infallibility, Big Blue's executives fell out of touch with the market as the action shifted

- from high-profit margin mainframes
- to commodity-like servers and desktop computers

Time and again the "infallible" IBM repetitively stumbled in product design and in product pricing. Its management's credo became the insistence that whatever it was that it offered *it had to be* appreciated by the market. There was no alternative. Customers were expected to behave.

In spite of one setback after another and a marked rejection that became more and more clear, by the mid- and late 1980s IBM's mainframe establishment never seemed to comprehend the extent to which the company's clout was fading. Yet, unlike its former competitors who were dumb enough not to want to change their tunnel vision centered on mainframes, companies marketing client-servers and database computers were making big inroads because of their high cost/effectiveness.

This insensitivity to client wishes and drives, as well as the resulting IBM troubles, raise questions about whether big companies can work anymore. In the late 1980s Big Blue was the case of a firm that used to be successful and fell on hard times. Its downfall provided lessons for other companies.

The mainframe legacy was perverse, and devil in information technology can still be found all over the world—particularly in Europe. The major impact has been both human and financial. Two generations of IT specialists and chief information officers (CIOs) have been trained and got their skills in

- mainframe chores,
- cobol programming, and
- very slow and inefficient software development timetables

Millions and millions of Cobol code have piled up and continue piling in the companies' programming libraries as well as in government agencies. The CIO of a major British bank was telling me that he hires young university graduates to maintain programs that were written before they were even born. He hates it, but he has no option.

Only the best managed companies used the year 2000 (Y2K) scare, of the late 1990s, to convert their libraries to C, C+, and other languages running on client-servers. The large majority still live under the mainframes and Cobol's yoke. New technologies come and go but the slavery is still there. The move to distributed processing was an opportunity to get rid of it. It did not happen that way. Students of the computer industry think that two reasons underpin this irrationality:

1. Human inertia
2. Invested interests

By late 2009 and early 2010, the new hope has been that *cloud computing* may be the liberator, but this is far from being sure. IBM is actively promoting "cloud computing mainframes," so called because the processing unit is still housed in a huge metal frame. And Big Blue rules still supreme in corporate data centers.* The PCMs challenge is something of the past and DEC faded away nearly two decades ago.

This policy of "backwards into the future" fits well with IBM's business strategy because many big companies still run crucial applications on the *big iron*. In fact, plenty of CIOs still like their mainframes because of IT's political and obsolete cultural reasons. They find as an excuse that they are "reliable" (which is nonsense), "secure" (an overstatement), and "easy

* There are still about 10,000 in use worldwide.

to maintain" (at great cost by Big Blue). But they are forgetting that mainframes and their Cobol suite are very costly, obsolete, unreliable,and labor intensive. The irony is that

- with cloud computing and mainframes these people and companies are living with one leg in the past and another in the future,
- but they do not appreciate that such a duality is a highly unstable balancing act; a mare's nest with a great lot of unexpected consequences

IBM, of course, has its reasons for trying to reestablish the mainframes virginity in the EDPers mind. Though they represent only 3.5 percent or so of its overall revenue, each dollar spent on hardware pulls in a much higher "I see dollars now" ratio from sales of software and maintenance contracts. Some experts estimate that 40 percent of IBM's profits are still *mainframe-related*.*

But not everything is so bright with this mainframe strategy. Aside the fact that not in all user organizations IT management is that dumb, quite a challenge is the run-in by startups like Neon. The software company sells a program allowing programming routines that run on mainframes to be shifted to client-servers under Linux, at great savings to the user.

Evidently the Big Blue does not like this intrusion in what it considers to be its secure domain, particularly now that it has practically got rid of all the other mainframers. It is said to have threatened to charge higher licensing fees to customers using Neon's solution†—which led Neon to file a lawsuit against IBM. If the court decides against that odious policy of arm-twisting and of monopolizing software on its mainframes, this would make a big dent in IBM's revenues—and in its prestige among its CIOs' disciples. We shall see.

---

* *The Economist*, January 16, 2010.

† Does it sound like IBM's late 1970s reaction to the PCMs? It does.

# Part Five

# Financial Staying Power

# 14

# *Financial Administration and the Budget*

## 14.1 FINANCIAL ADMINISTRATION

Financial administration has many aspects, three of which will interest *us* in this chapter: financial plan, budget, and cash management. The financial plan is developed as a crucial question confronting every enterprise: From where will come the money for going from "here" to "there?" Any business operation requires financial means, not only the human resources intended to see it through (see also Chapter 16 on financial planning and control).

The board and CEO may have decided upon an objective, but does the company have the financial staying power to achieve it? Financial planning is not only a question of money, because money may be awfully misused without a skilled and well-structured administrative support. Therefore, to thoroughly understand financial administration, and its related functions, several crucial factors must be considered:

- The very scope of business activities
- Structural elements connected to action
- Cash inflow and outflow
- Administrative systems and procedures
- Experience of the personnel
- Efficiency of administrative controls
- Methods, facilities, and equipment

The needs for well-tuned financial and accounting services is present all the way from R&D to production and sales—and this not only at headquarters but also at all of the subsidiaries. Auditing has to check financial administrative functions, even those considered to be mainly routine and limited in scope, because they may hide dragons.

Within this broader perspective, financial administration starts with financial planning long, medium, and short range. The budget is 1-year finance plan (Section 14.4). Other financial administration duties are the authorization of expenditures; plan versus actual comparisons; corrective measures; and the preparation of financial reports for management as well as legally required quarterly and annual financial statements.

Even the best organized financial administration chores decay with time. Therefore, everyone of these duties must be critically examined in regard to its accuracy and effectiveness. Also, in regard to the way in which financial information is brought to senior management's attention.

Decades of experience in top management consulting tell me that most financial reports reaching the vertex of the organization are collections of data some of which are interesting and others dull. But they are rarely sufficiently concise or organized to provide management with the significant facts it needs to plan and control the company's finances. In one investigation, I found 87 reports regularly directed to the president's office; their origin and frequency is shown in Table 14.1. Analysis revealed that much of the information included in the reports

- was duplicated because they were prepared independently in different areas of the company, and
- about one-third of vital references the president needed to effectively direct financial matters was utterly absent

As if this was not enough, about a quarter of data reaching the hands of top management was frequently inaccurate, or incompatible with the type of the report identified in the headline. A typical example is the quarterly analysis of product costs, normally released some 40 days after the completion of the quarter. At that point effective corrective action can no more be initiated.

These fault lines are not found only in the reporting structure. In many cases, the Financial Administration's organization includes functions not essential to planning, execution, and control activities. The inclusion of reports regarding details of receiving, storage, and shipping business if

**TABLE 14.1**

Origin and Frequency of Reports Submitted to the President

| Source | Periodic | Daily | Weekly | Bimonthly | Monthly | Quarterly | Annual | Total |
|---|---|---|---|---|---|---|---|---|
| General Accounting | 1 | | | 1 | 5 | 17 | 1 | 25 |
| Sales | | 1 | | | 14 | 2 | 7 | 24 |
| Public Relations | | | | | 1 | | | 1 |
| Production Planning | 2 | | | | 1 | | 2 | 5 |
| Manufacturing Management | | | | | 2 | | | 2 |
| Industrial Accounting | | 1 | 1 | | 8 | 1 | | 11 |
| Stores & Transportation | | 3 | | | | | | 3 |
| Technical | 8 | | | | | | | 8 |
| Purchasing | 1 | | | | 1 | | | 2 |
| Personnel | | | | | 4 | | | 4 |
| Auditing | | | | | 2 | | | 2 |
| **Totals** | **12** | **5** | **1** | **1** | **38** | **20** | **10** | **87** |

often justified as an attempt to control the physical movement of materials. Yet, the same results can be achieved through

- an integrated system of control documents, and
- a periodic audit of actual conditions to ascertain that what is reported is accurate

Such an audit should include a check on adherence to authorized procedural practices and a sample-based physical check on inventory conditions. The result of mixing planning and control with actual operations sees to it that in the end many of the functions performed by the division are routine and a distraction from its main duties—which essentially are financial plans, checks, and balances. When a tremendous amount of raw data passes through

- this information is seldom screened and analyzed for control purposes, and
- it rarely does ever reach the hands of individuals directly responsible for specific duties, such as controlling costs

*If* headquarters operates under these conditions *then*, by extension, controls over branch operations are limited and ineffective. Moreover, in several companies factories and sales office operations are run as independent businesses, each with its own financial force. The principal direction seems to come from the Manufacturing or Sales division, while administrative supervision is limited to a series of routine and somewhat ineffective reports.

A case that I had years ago still remains etched in my mind. It concerned controls over spare parts inventories that were highly ineffective, demonstrated by the low average turnover rate of under two times per year. The cost control audit I did established that inventories were kept at unreasonable (and unprofitable) high levels, of stock as shown in Figure 14.1. No targets or objectives were established in order to control inventory levels.

A similar reference is valid in regard to cost control (Chapter 15). While average monthly cost records were maintained, no follow-up was made to guide production operations in curbing cost increases. Neither were factory managers held accountable for what appeared to be persistently above-average costs.

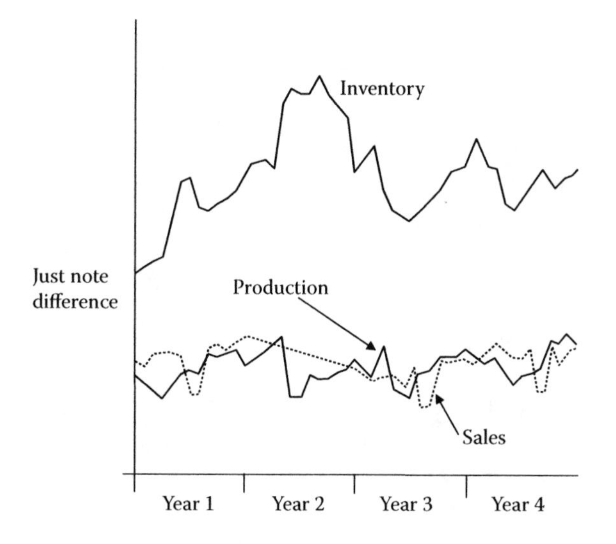


**FIGURE 14.1**
Inventory levels of a general purpose product were out of synchronization with production and sales.

The reader should notice that this error is widespread and often finds its origin in the fact that in many cases executive responsibilities are not being clearly defined or understood, with the result that one manager pouts sticks in the wheels of the others. In others cases, high costs are regarded as an expedient to meet production schedules.

A common weakness underpinning all this is that below the first line of supervision there is little depth to the organization. The absence of qualified young people throughout the Financial Administration division is also often noticeable. Training qualified individuals to take over key positions is rarely underway, and then usually limited to old-fashioned accounting procedures.

## 14.2 A GENERAL ELECTRIC CASE STUDY ON FINANCIAL MANAGEMENT

At practically all levels and functions of management, few people have the experience and intelligence to appreciate that they consistently overestimate the information available to them for their decisions. Another administrative ill is a curious belief in certainty, whether this has to do with historical inevitability; different grand designs, or the firm's financial staying power.

Quite often forecasts, particularly those long-range, are no better than numbers pulled out of a hat. Peter Drucker once advised that the best contribution of a forecast is that it offers a way to judge the future impact of present day decisions; provided, of course, that these decisions have been properly studied and benefited from thorough evaluation of alternatives.

In no other domain is this more applicable than in financial matters, if for no other reason because financial issues underpin all others and are therefore omnipresent.

Precisely for this reason, since the early decades of the twentieth century, GE adopted a clear-cut division between operations and finance. The latter's remit is planning, such as budgeting, as well as making short-, medium-, and longer- range financial evaluations, and providing feedback—which is an important control function.

The general manager of a GE division or subsidiary abroad is responsible for R&D, production, sales, services—in short all activities except finance. The director of finance reports through the financial hierarchical line of authority, which goes all the way up the ladder to the EVP finance at headquarters—who is also responsible for his or her salary and promotion.

Another important principle from GE financial management, which can be found as well in DuPont and GM—is that while the budget is the annual financial plan it provides no automatic authorization to spend money. Expenses must fit within the budget, but they must also be individually authorized.

A great GE financial management principle is, as well, the responsibilization of each person in position of authority in regard to self-analyzing his or her performance. When he was at the helm of GE, John F. Welch, Jr. required his executives to analyze themselves and their operations, comparing their deliverables with those of competitors around the globe. His theory was that in today's brutally competitive world

- size was no longer the trump card it once was, and
- the role played by size in the past has been taken up by innovation and the ability to capitalize on deregulation and globalization

Although the scale of big companies sees to it that their reach seems to impress, the enterprises that more frequently create wealth are those who thrive on imagination and (as Welch stated), ridicule bureaucracy but are alert to customer demand. Analytical finance is so important because through it management faces the reality every day, and when it has to move, it cannot do so with speed.

Integral part of analytical finance is, however, *risk control*. Sound financial management looks at creative accounting and at the taking of risks without checks and limits as anathema. Most certainly this is not what should be done,* and General Electric Capital provides an example on what can go wrong (more on this later).

Good news first, Welch has been instrumental in overseeing GE's elimination of boundaries, fiefdoms, and bureaucracy in an effort to create a boundaryless organization. Among the payoffs were sharp reductions in the time it takes to deliver products from the moment they are ordered. (For instance, by 1996, GE Appliances was able to deliver in 3½ weeks, down from 18 weeks in 1990. In the ensuing years this was reduced to three days.)

Fast delivery means fast billing and better cash flow. In addition, speed of delivery is a big challenge to the organization as a whole keeping everyone on the run.

A company that has been managed in the old, classical, slow moving, bureaucratic way, Jack Welch emphasized, is one going downhill. The prototypes of its managers are Roger Smith of GM, and John Akers of IBM. Typically, CEOs unwilling or unable to effect the dramatic change necessary to prevent a debilitating decline in the company's fortunes would

- continue with the old product line(s), insensitive to market's switch,
- guarantee the employment of tens of thousands of employees the company no longer needs, and
- fund unlimited R&D mostly in useless projects, rather than aggressively expanding in sectors that would bring a future bonanza

At the early years of Welch's tenure, GE was a company accustomed to design and sell jet engines, light bulbs, power production equipment, and home appliances. To diversify, he developed a successful finance arm: GE Capital Services, which became the engine driving its parent's profits (but hit the skids with the 2007–2010 deep economic and financial crisis).

* Founded in 1932 as General Electric Contracts Corporation to provide financing that supported the firm's industrial businesses, the operation gradually expanded into other areas of lending unrelated to GE.

GE Capital's portfolio included nearly two dozen discrete businesses and extended to almost every corner of the world of finance. It pushed ahead globally, while its association with GE allowed it to borrow money cheaply thanks to the parent company's AAA credit rating. (But a major downturn can wreak havoc with credit rating, not only of companies but also of sovereigns.) Welch instilled

- a drive for performance,
- an appetite for risk-taking and deal-making,
- the ability to shift strategies rapidly to take advantage of change, and
- an engineer's ability to run operations, not just write a check to finance them

On paper, all that looks great, but somehow financial management did not stand up to the job, because risk management was weak—particularly in the side of GE Capital. Under Jack Welch, GE Capital grew rapidly, but it has also been largely obsessed with short-term profit goals.*

If GE Capital were a bank, it would rank as one of the biggest in America, nearly at par with Morgan Stanley and way ahead of U.S. Bancorp and Bank of New York Mellon. "GE used GE Capital like a cookie jar" into which it dipped when needed, says James Schrager, a professor at the University of Chicago's Booth School of Business.†

GE Capital flourished as a member of the so-called shadow banking system; companies that offered a golden horde of financial products without having to bear the regulatory burdens of banks. But risk management was substandard and as GE Capital's credit rating continued to slip, it raised its funding costs and obliged its parent to come up with lots of funds to shore up its subsidiary.

In early September 2008, at the time of the Lehman bankruptcy, the spread of GE's five-year CDS hit 600, the largest ever for AAA rated companies. In February 10, 2009 GE shifted $9.5 billion to beef up GE Capital finances and improve its debt ratio. A few weeks later on March 3, 2009 GE's stock dived to below $7 and bets were made that it would go down to $2.5. It did not, but the scare at the company's top was unprecedented; poor risk management had led the firm close to the abyss.

* Although Welch has recently argued that it is a "dumb idea" for managers to become obsessed with short-term profit goals.

† *The Economist*, March 21, 2009.

## 14.3 IMPROVING FINANCIAL PERFORMANCE THROUGH DIVERSIFICATION: AMADEO GIANNINI

Financial performance can be improved in two ways. The one is better management assisted by a real-time, interactive planning and control methodology. The other is a system of adjustable limits, able to assure that risks being taken are well established, measurable, and steadily controlled. Most definitely these should include limits to debt and overhead expenses (respectively Sections 16.2 and 16.3 in Chapter 16).

Amadeo Giannini, the man who made Bank of America and Transamerica Corporation, provides an example on how better management keeps exposure under lock and key. He is widely credited for having pioneered home mortgages, auto loans, and other installment credit; but he never went overboard in his innovations.

Giannini has also been the architect of what has become nationwide banking, even if parochial interests prevented him from realizing it in its life-time. Many decades later, the first bank in the United States to have branches coast-to-coast has been Bank of America, which accomplished the feat through its $48 billion merger with NationsBank of Charlotte, North Carolina.*

One of Giannini's guiding principles was that there is money to be made by lending to the little fellow. He promoted deposits and loans by ringing doorbells and buttonholing people on the street, painstakingly explaining what a bank does. It should be remembered that, at the time, it was not considered ethical to solicit banking business.

Amadeo Giannini also made a career by lending to out-of-favor industries. He helped the California wine industry get started, then bankrolled Hollywood at a time when the movie industry was far from being proven. Behind all this was his vision of *spreading risk through diversification*. Other things equal, a bank doing business in different industries and in all parts of a state or nation

- would be less vulnerable to any one of industry's or region's difficulties, and
- it would be strong enough to continue lending to troubled communities, when they were most in need

* Roughly midway between Giannini's concept and coast-to-coast banking, in 1958 Bank of America pioneered BankAmericard (now Visa), the first widely used credit card.

This has been one of the best examples of the business of banking as an intermediary, but Giannini also put in place risk controls—one of the most powerful being personal knowledge of debtor rather than depending on the vague and inaccurate credit ratings of independent companies. (A policy also followed by Andrew Mellon, another great banker of the early twentieth century.)

The financial management principles outlined in the above paragraphs are guideposts to sound governance. Say, as an example, that the management of a typical company expresses the desire to improve performance and increase the stature of existing operations to better serve the needs of management. To accomplish this, it is betting on seven management milestones observed by nearly all great entrepreneurs:

- Long-range planning (Chapter 4), by developing a company-wide plan of action within established objectives.
- Effective financial planning, elaborating a shorter-term plan able to maintain adequate capital for established financial objectives.
- Budgeting (Section 14.4), spelling out definite operational goals that become the guiding posts covered by adequate finances.
- Profit analysis, based on a continuing review of income, costs, and profits to ensure the firm's financial staying power.
- Cost/effective inventory control, the development of inventory targets, and the timely control of their status.
- Cost control, by the steady accumulation, analysis, interpretation, and reporting of detailed unit cost data.
- Manpower control, based on regular analysis, interpretation, and reporting of efficient utilization of human resources and avoidance of idle time.

Beyond these basic elements to be incorporated into the financial planning and control system, there are other refinements that must be considered and adopted—including those concerning the product line and marketing strategy as well as capital allocation and expected ROI.

Effective capital allocation depends to a significant degree, though not exclusively, on projected profitability of investments associated to lower costs and growth targets. A thorough evaluation requires the testing of different hypotheses. Furthermore, to establish the amount of money to be invested, it is necessary to translate plans into specific sales objectives.

A consolidation of the individual plans by operating division results in an overall profit plan for the coming period; this being followed by means of feedback that compare actual to planned results, and help in determining reasons for variations. Feedbacks and management reports based on them must be designed in a way to induce immediate actions. They must be specific and pertinent to the job of the functional executives receiving them. The following are basic topics for senior management reports:

- Cash flow statistics over a five-year period, and next period projections
- Incurred expenses (total, cost of goods sold, and cost of sales) versus those planned
- Profit and loss (by operating division, branch, subsidiary, product line, product, and the company as a whole)
- Actual versus planned turnover (as above)
- Actual versus planned expenses for R&D (by project, lab, and the company)
- Actual versus planned manufacturing costs (by factory, product, process, country, and area)
- Analysis of shipments (from any place to any place) and of backlogs
- Analysis of business lost by country, area, and product—including reasons why business was lost

Apart from these analytical results to be presented in a factual and documented way, management reporting should focus on detecting errors, omissions, inconsistencies, or a lack of realism in the budget. The association of merits and demerits to good and bad performance, respectively, serves in improving outgoing quality.

In the late 1970s, Citibank dedicated a DEC PDP 11 to tracking the quality of documentary credit (financing of export business). Every time a customer order was executed, it was registered on the minicomputer and followed up. If there was no claiming of an error by a customer, its author was credited with merits. To errors corresponded demerits, and

- the system was giving a deadline within which the error had to be corrected,
- if not, more demerits followed, which found their way to the personal file of the faulty order's author

Plenty of other applications that are important to sound management can benefit from individual tracking. An example is low probability but high impact events, such as Iceland's volcanic eruption, and their effect on the bottomline. Grounded planes don't earn money, while obligations taken by airlines tend to rapidly increase. As the volcanic cloud spread over Europe, for several days airports have been closed and the estimated cost to the airlines reached € 1.5 billion (then $1.9 billion).

It is not only volcanoes that create havoc to company finances. In a technologically advanced economy software failures and system downtime at large can be lethal. On November 19, 2009, a computer glitch caused widespread flight delays and cancellations in the United States.

- The system was practically down for four hours.
- The failure has been attributed to software connected to the infrastructure within the Federal Aviation Authority (FAA).

A news item at Bloomberg television called it "the 20 billion dollar glitch." That's an excellent example of risk connected to high technology, a domain that is frequently taken as free of exposure. Financial Administration must fully account for both the likelihood and the impact of tail events. If it does not, a company (and for that matter a nation) would undergo a period of financial stress for which it is not at all prepared.

## 14.4 THE BUDGET

The process of financial planning contributes directly and most significantly to effective management, provided that certain conditions are met. One of them is that each of the company's primary functions is directly served by *budgeting*, based on careful study, investigation, and research undertaken in order to determine expected future operations and associated costs. Therefore a budget would contain sound, attainable goals.

It is precisely because of this that *budget* is a formal written statement of management's financial plan for the future, expressed in quantitative terms. The allocations of funds to budgetary chapters, and the statistics being derived from the way such allocation is used, chart the course of future action and provides a basis for control. The prerequisite is that the budget contains sound, attainable objectives rather than mere wishful thinking. When these conditions are satisfied,

- the budget becomes a *planning model*, and the means used for its development are the planning instruments, and
- such instruments work and provide a reliable output well when the company has the proper methodology and solid cost data

The reason why a great deal of attention must be paid to costs is that the whole process of financial planning is based on them. In the process of costing (Chapter 15), the budget becomes an orderly presentation of projected activity for the next financial year, based on the amount of work to be done. Typically, an industrial company's budget divides into four main chapters:

1. Research and development
2. Production schedules
3. Marketing and sales
4. Administrative activities

Each is divided into work elements reflecting operating expenses (opex). Multiplied by the corresponding costs, such expense areas condition the level of budgetary allocation in direct and indirect expenses. The budget also includes other chapters, such as capital investments (capex), and has associated to it a cash flow plan (Section 14.6).

The planning premises entering a budgetary process increase management's ability to rely on fact finding, lessening the role of hunches and intuition in running the enterprise. Organization-wide coordination through budgetary procedures is facilitated when each level of management participates in the preparation of the budget.

- Top management should be setting and explaining objectives,
- but each organizational level must establish its budget under such guidelines, subject to subsequent approval.

Modeling is important, but there also exist other prerequisites. One of the crucial issues with budgeting is the input of information to the database, necessary to permit realistic financial plans. The need for accuracy of both *unit costs* and *activity projections* must be fully appreciated. Some of the input will come

1. *top down* from senior management to the divisions, departments, operating units, and back to top management, or
2. *bottom up*, elaborated at operating unit level, from there send to the department, the division and to top management where it gets approved and consolidated

Wherever both channels are feasible, the No. 2 is far preferable since it leads to a financial plan, which has the salt of the Earth. It starts close to the lower management level where the budget will have to be subsequently applied, and therefore it motivates the people who, later on, will have to work with that budget.

Various options are available to facilitate the input task to the different budgetary chapters. Many cost features, for example, vary only slightly from year to year and their automation enables managers to concentrate their efforts on those of their accounts that do warrant careful analysis. Short-cut methods include

- quarterly and monthly amount to be prorated according to the actual number of days in each period

The aim is to have projected amounts created automatically by the system, using forecasting techniques and standard costs.

- A properly defined increase (or decrease) over the current year's actual and projected basic amounts, to provide a basis for alternative budgets depending on economic activity

At times such entries may be coarse and will need to be subsequently refined, but as they stand they do constitute starting points, and provide a frame of reference for evaluation.

- Combination entries, with distributions resting on algorithms, which map marked variations and or seasonal fluctuations

The models used to produce this information also help to control plan versus actual results (Section 14.3). Moreover, a priori profit and loss evaluations can be based on information derived from the financial plans and operating statistics—producing ad hoc interactive reports from cost factors to projected customer profitability.

As a result of this budgetary process, the activities of each department are integrated with those of related departments in the same area, nationwide and internationally. But as it has been underlined from Section 14.1 though managerial planning and coordination are important, they must be always accompanied by control.

- Budgeting contributes to effective management control by providing the standard against which actual performance will be evaluated and variances revealed.
- Actual versus plan evaluations need to be done for each budgeted position, operating unit, and the company as a whole; hence in both a detailed and a consolidated way.

The accounting system should be company-wide homogeneous and integrated. In terms of financial information, no divisional accounting is an island. To function properly, all units and their budgets have to work in synergy. For instance, the purchase order subsystem works in conjunction with the inventory management, production planning, and accounts payable subsystems. The model that we develop should pay due attention to this fact.

An integrative view of accounts is necessary, starting at system design level. The budgetary methodology that we establish must be mapped through software in a way which ensures that the different financial subsystems interact and relate among themselves. This is written in the understanding that a flexible but integrated budgetary application is of critical importance to the profits and competitiveness of the company.

Furthermore, the solution we adopt must be able to smoothly transfer the most up-to-date information from module to module, and from there to the end-user. While this reference goes beyond what is generally considered to be the budget's remit, it is one of the pillars of budgetary control and therefore of sound corporate management.

## 14.5 THE INTEREST AND NONINTEREST BUDGET

Provided we have properly elaborated and tested the model that we employ, budgeting helps to establish a series of relationships between activities across the enterprise. In a financial planning sense, the actual relationships or linkages between individual elements determine the structure of the model. Under this perspective, the better approaches to budgeting tend to share two basic characteristics:

- A definition or statement of which elements connect to one another, and
- An explanation of the manner in which the elements comprising the model are related among themselves

That much about the dynamics. The mechanics vary from one firm to another because not all budgeting solutions are exactly the same. Differences exist particularly in the flexibility with which budget updates and modifications can be made, as well as the accuracy characterizing the financial plans themselves. With greater flexibility, it is possible to customize budgetary information by linking together a series of selected accounts in an ad hoc manner. This can be done, for instance for

- specific end-user whose job is improved through personalization, and
- profit center or cost center reporting (Chapter 15), where focus is at a premium

Recorded in cash flow accounting (Section 14.6), the actual financial budget transactions can be used as a basis for actual versus plan comparison. Critical evaluations can be made along a chosen line of managerial importance, such as profitability, cost overruns, and other budgetary evaluation criteria.

The outcome of analytical procedures enables us not only to control current expenditures but also to prepare more accurately the new budget that can be checked through *variance analysis*. This leads to the issue of both algorithmic and heuristic support for financial planning and budgeting, as well as for purposes of financial control.

The specialization of budgets often helps in providing a more clear picture of commitments. An example is provided by the financial industry, where a major distinction is made between the *interest* and the *noninterest* budget. The two share total capital outlays at the approximate ratio of 2:1; about 66 percent of a bank's annual expenses go to the interest budget and 34 percent to the noninterest budget. These are two clearly distinct classes of commitments, each requiring an appropriate model.

- The interest budget covers the money paid to depositors, other banks (bought money), the reserve bank and other sources—that is, the *cost of money*.
- The noninterest budget addresses *all other expenditures*: personnel, real estate, utilities, IT, communications, and so on. This is close to the budgetary notion that prevails in the manufacturing industry.

Every financial institution must know and control each of its noninterest expenses from salaries and wage benefits to investments in high technology (communications, computers, software); brick and mortar (headquarters, branches) and associated occupancy costs, as well as other running expenses.

Being careful about money spent on the interest budget involves the critical evaluation of the cost of deposits (including part of the cost of the branch network), against the cost of bought money. Such an evaluation also brings into the picture issues connected to the institution's name and creditworthiness, which go beyond pure costs. During the first couple of years of the 2007–2010 deep economic and banking crisis, the interbank market had dried up because credit institutions no more trusted one another.

The reader should as well know that while one of basic goals of budgeting is to facilitate analysis, there is no standardization regarding the items that come under interest and noninterest expenses (even from one to another financial institution there are no major differences either). In fact, some differences are only in terminology. Some banks use Net Interest Income. Others favor Net Interest Revenue or Net Interest Earned.

Quite similarly, several banks use the term Net Interest Margin; others prefer Net Interest Spread: Net Margin; Interest Differential, or Effective Interest Differential. More serious is the fact that there are no universal standards for the monitoring of net interest margins and other variables—or for that matter standard reporting periods.

- Generally, in the banking industry the typical margin reports are for periods covering six months to a year.
- But leading financial institutions suggest that waiting so long for the evaluation of vital information does not show quality of management; reporting should be much more frequent.

In managing the exposure that the bank is taking in the side of the interest budget, the executives and professionals of the institution should *at any time* be in a position to reach online into databases in order to

- critically evaluate conditions of imbalance, and
- judge the volatility in interest spreads and their aftereffect on the institution

Provided the database is properly designed, updated, accessed, and available online, expert systems can be used in an effective manner to help the management of the bank in its mission to control margin imbalance and volatility. Knowledge engineering artifacts should be used to focus the executive's attention

- on deviations from standards (and/or trends), and
- the reasons underpinning such deviations

Interactively available on request, analytical reports should rapidly pinpoint change as well as exceptions. In this way, management can act while there is still time to favorably affect the profit and loss statement. This is true both of the interest and of the noninterest budgets; generally, of any commitment the bank has made. For instance, issues requiring real-time attention in regard to exposure are typically connected to

- net interest margin
- total earning assets
- investment portfolio
- loans portfolio
- collateralized debt obligations (CDOs)
- other securitized instruments, and
- credit default swaps (CDSs)

Advanced mathematical tools are not only welcome but also necessary. Because interest rates fluctuate, the cost of funds varies significantly and no deterministic standards procedure provides a longer term basis for management accounting. By contrast, a good tool is *fuzzy engineering*, which permits to handle the vagueness and uncertainty of interest rates and other economic variables in an effective manner.*

* D. N. Chorafas, *Risk Management Technology in Financial Services*, Elsevier, London and Boston, 2007.

The solution to be adopted should see to it that interactive financial reporting presents both summary and (or request) detailed information on each component of earning assets, other assets, liabilities, and equity. Changes in margin from the previous period must not only be included but also explained. This is a good example of plan versus actual evaluation in a financial environment, which includes significant uncertainties.

Focused, timely financial information allows management to quickly monitor changes, subsequently taking corrective action. The message these paragraphs convey is that the budgetary process is not a one-way street just piling up expense after expense and asking for its funding. By contrast, it is a *two-way process* that can only then be executed when there is a total view of

- planning premises,
- execution of operations, and
- control procedures

A sound way to look at requirements for analytical finance is to examine them in a multidimensional way by organizational unit, product or service, customer relationship production and distribution facilities, aligning the plan with the way actual results are reported (and vice versa). Whether for planning purposes or for management control reasons, the online interactive presentation of budget information should be *future-oriented.*

## 14.6 CASH FLOW

A good cash flow means *financial staying power.* In the longer term, it is a function of *our* products and *their* market appeal, which rest on the imagination of R&D engineers, and the ingenuity of the salesmen. Cash flow is an important criterion in financial analysis because in the very short term (90 to 180 days), it makes the difference between the company's solvency and insolvency—while in the longer term it contributes to its financial freedom, from loans and indebtedness.

Cash flow is typically generated through business transactions, and we all know that such transactions are of a *circular* character. This cyclical process begins with a *cash debit* and ends with a *cash credit*—which creates the basis of the cash flow and its computation.

One of the problems with *cash flow management* is that the term does not have just one definition and method of computation, as some textbooks suggest. Much depends on the type of company and the accounting practices that it has adopted. Therefore, in order to remove ambiguity it is wise to start with the fundamentals: more precisely, with the meaning of *cash*.

Cash is a liquid asset, and it assumes many forms. There is cash in the bank, cash on hand, undeposited checks and money orders, undeposited drafts, change funds, payroll funds, and so on. In the balance sheet cash is typically classified as either

- cash on hand, or
- cash in the bank

Cash on hand includes coins, paper money, checks, bank drafts, cashier's checks, express money orders, and postal money orders. Cash in bank consists of demand deposits in checking accounts. Demand certificates of deposit may be classified as cash, but this is not true of time deposit certificates. Safeguarding the cash is only one element of a good system of internal checks and balances, but it is a very important one. Adequate internal check upon cash require steady control over both *receipts* and *disbursements*. Both require careful tracking. One of the uses of cash flow is to face the demands posed by accounts payable, which as their name implies, are the sums owed for goods and services bought and charged.

Notes payable are disbursements. They represent promissory notes owed to the banks or to a financing company. The cash flow also serves to pay interest and capital on bank loans, as well as on *bonds* that can be seen as a longer term loan given by the capital market.

Typically, bonds are issued by creditworthy companies and they are floated at a given coupon rate. Bullet bonds (straight bonds) are not due prior to maturity, but this is not true of callable bonds. If the credit rating of the firm is low, dividends and capital due for high interest bonds (junk bonds) consume a significant amount of cash flow. As these examples demonstrate,

- the diversity of financial obligation to be faced through the company's cash flow requires very careful study and experimentation, and
- both simulation and knowledge engineering can be of significant assistance to the financial analyst, and to the company's management as well

Cash flow estimates are not made *one tantum*. They must be carefully evaluated, reevaluated, and controlled. This is a never-ending job. The methods to do so vary greatly between organizations, and the same is true of financial management at large.

In its simplest form, *cash flows* is net income plus items such as depreciation. *Net income* is both the bottom line and the starting point when figuring out a company's cash flow. The algorithm is as follows:

Cash Flow = Net Income + Depreciation + Depletion + Amortization

In current industrial accounting practice, *depreciation* costs are subtracted from net income. This is seen as a major reason for the depressed profits of many companies. But it also means that much of the cash that would otherwise be invisible is brought under perspective.

- A company that depreciates $100 million shows net earnings of another $100 million and pays no dividend
- has actually generated a cash flow of $200 million, twice its earnings

The concept of *depletion* is the write-off when the asset being used is no more available. Assets of natural resources, such as oil, coal, or minerals deposits are depleted. *Amortization* is also a write-down but of specific terms applying to acquired assets such as land, factories, machinery, or of intangible assets.

Simple cash flow is not a particularly useful figure by itself, what is important is *operating cash flow* (OCF). It denotes the money generated by a company before cost of financing and taxes come into play.

OCF = Operating cash flow
= Cash flow + Interest expense + Income tax expense

*Interest expense* is added back to the simple cash flow, to get the broadest possible measure. In case of takeovers, *income tax expense* is added because it will not have to be paid after the new owner adds so much debt that there is no book profit and hence no tax due.

Another important money indicator to watch is *free cash flow* (FCF); it tells how much cash is *uncommitted* and available for other uses. Its computation takes the cash flow figures and adjusts it to account for certain balance sheet items. For instance, subtracting current debt and capital expenditures.

FCF = Free cash flow = Cash flow – Capital expenditures – Dividends

A very important metric connected to cash flow estimates is *net present value* (NPV). I look at it as the most important single value measure of cash flows, adjusted for *risk* and the *cost* of money. Few companies are properly tooled to take advantage of NPV because its computation is not algorithmic. It involves subjective judgment in connection to *discounted cash flow*, which requires

- well-documented timing assumptions, and
- a realistic rate for cash flow discounting (hurdle rate)

More precisely, the internal rate of return (IRR) or the rate at which the net present value equals zero. A standards present value calculation, assuming noncontinuous discounting, can be done through models. A similar approach is valid for *asset recovery* with discrimination between assets that were sold on the first day of the year (because their useful life had ended) and assets sold on the last day of the year and/or some other timeframe (maybe for working capital liquidation).

Net present value can as well be computed by *marking to market*. The problem is that not all assets, and surely not all financial instruments, have an active market. Guesstimates are made through the use of models, but many models are asymmetric and sometimes marking to model is equal to marking to myth.

As these references document, while the measurement of cash in coins and banknotes is a relatively simple issue whose valuation depends on arithmetic calculation, other metrics of financial wealth, like net present value of current assets, represent a most challenging subject. It therefore comes as no surprise that the proper management of cash flow in its various definitions has been a concern of financiers, merchants, and industrialists for all of recorded history.

## 14.7 CASH FLOW MANAGEMENT

The proper measurement and management of cash flow is so important because its presence reflects the company's ability to continue operating as a going concern. It follows that the absence of policies and procedures to assure a sound cash flow has been a critical as well as recurrent theme in business. Not every executive realizes that, in its fundamentals, the challenge of projecting, measuring, and controlling his company's cash flow is critical to continuing performance, and it essentially amounts to aggregating financial resources with particular emphasis on their evolution as time goes on. The crucial test is one of taking all business parameters into account in order to develop factual and documented cash flow scenarios. Just as important is to appreciate that

- there is a close connection between budgeting, cash flow, and risk control, and
- forecasts have to be made for cash flows *prior* to new commitments, including the evaluation of financial impact from commitments already made

The concept behind these two bullets extends the budget/cash flow duality all the way to the balance sheet, where they are both integrated. We talk, of course, of truthful balance sheets and not of double and triple books. For instance, balance sheets are false when

- the assets are overvalued,
- the liabilities are underpriced,
- they are based on "proforma" and refer to unaudited or improperly audited financial statements

What every management worth its salt should be careful to avoid is *creative accounting.* Lies have short legs, and it serves precious little to lie to ourselves and to the market. Yet, creative accounting is widespread and, as such, it poses questions about the

- level of ethical standards, and
- the quality of business management

Contrary to creative accounting, which is systematic, errors can happen now and then in financial calculations. Cash flow estimates may prove inaccurate because their resources have been lightly researched, rest on overoptimistic assumptions, or the likelihood of major events has been foolishly discarded.

The crashing of fresh orders for the world's shipyards provides an example. In 2009 they were more than 80 percent lower than two years earlier when, prior to the July 2007 crash, high freight rates and rosy economic forecasts encouraged shipping companies to scramble for new vessels. The subsequent recession sent shipping rates tumbling and this reverberated all the way to the shipyards' cash flow.

Statistics focusing only on one of the pertinent factors can be misleading. For instance, in the first quarter of 2010 South Korea's shipyards won over half of global orders for new ships. But they were worth just $2.2 billion, while in 2008 Korean shipyards won orders worth $32 billion. This meant dwindling cash flows that hit companies in two ways:

1. They diminish their financial independence.
2. They make it more difficult for them to obtain loans, as the better managed banks are carefully watching debtors for brewing troubles.

Therefore, cash flow forecasts should be elaborated with great care. They should definitely cover at least rolling year period—including accounts receivable and to be paid during *this* timeframe, plus other income. For instance, patents and other accords on know-how, portfolio holdings, sales of property, and so on.

Management should spare no attention to detail, including *resource-leveling* exercises. This is a process of modifying the schedule of company activities, in order to reduce any peaks in financial requirements that cannot be matched through cash flow. Nobody can lose sight of financial staying power and prosper. Resource-leveling schedules are known from project management. They rest on two leveling algorithms that should be part of cash flow management:

1. Time-limited leveling
2. Resource-limited leveling

In resource-limited leveling the known capacity, or availability level, of each resource is never to be exceeded by the schedule. Activities are delayed until the financial means are available, with particular attention paid to not exceeding the availability level.

In time-limited leveling, activities are delayed only if they have float and if the completion date of the project is not affected. Time-limited leveling will reduce the peaks in the financial resource profiles, but it is not guaranteed to reduce the peaks to a level below known and relatively unknown factors affecting the cash flow.

A sound cash flow management is also promoted by a discovery process, aiming to unearth hidden factors. This is particularly important with new ventures, new structures, and new instruments. In the first few years following the massive creation of bank holding companies, back in 1970–1971, problems in bank analysis associated with cash flows were masked by the

- issues connected with the fully consolidated statements at holding company level, and
- widespread belief that growth would take care of those who worried about cash flow by individual units, or at the holding

The facts have shown that this is not true. The case of Drexel Burnham Lambert provides an example. The cash available at bank holding companies, as well as of their profitable subsidiaries, must do more work than pay for dividends to shareholders and service double-leveraged debt. There is a time asymmetry to account for in terms of financial staying power:

- Commitments regarding capital investments are made at the beginning of the year,
- but operating flows, representing revenues and expenses, occur throughout the year.

Hence an important mission of financial management is to study the characteristic pattern of the operating cash flow. To raise cash, capital assets specified by management are sold according to market opportunity within a given timetable; depreciated items with, say, a 6- or 7-year life-cycle, are sold at the beginning of the year following the year they were purchased; and disposable inventory is liquidated on the last day of the year, but taxes on its sale are treated as operating cash flows.

In conclusion, successful business organizations pay significant attention to managing their cash flow. They do so through analytics supported by advanced technology. A crucial question management should ask itself is, "Do we have the skill and the tools to take advantage of the opportunities propelled through the market forces? If not, what should we be doing to significantly improve our cash flow management?"

# 15

# *Profit Centers, Cost Control, and Standard Costs*

## 15.1 PROFIT CENTERS AND COST CENTERS

A *profit center* is an organizational unit whose manager has been given full responsibility to operate on a profit-and-loss basis, his performance being judged on results. As its name implies, a profit center's income statement must show steady profits, not losses. Subsidized profit centers are an oxymoron.

By contrast, a *cost center* typically offers support functions to other company departments and operates on a budget granted by general management. The information system department, for example, is quite often a cost center (which constitutes a very bad policy). Cost centers have neither a soul to blame nor a body to kick. Service departments should bill for their services because, inter alia

- this keeps them slim and fast reacting,
- the cost/effectiveness of their services is judged by the user departments who pay their bills, and
- billing turns a cost center into a profit center

The best way to look at *profit centers* is as the company's income earners. As such, they obey a pricing structure that should be neither abstract nor outside a factual range of market prices for same quality products. In the profit center restructuring I did some years ago in a credit institution, income from services was credited to each department according to its

- deposits
- loans

- trading, and
- wealth management

This restructuring did away with all internal cost centers. Providers of services, like IT, whose produce was oriented toward other units of the credit institution, had to bill for their services using competitive prices. They also had to project on future income, cash flow, (Chapter 14) and expenditures, and make plan versus actual evaluations that went straight to management accounting.

These were new experiences to the bank, and at the beginning they complained. But the CEO did not accept turning back the clock, and after some resistance they fell in line. Particular attention was paid in the course of this reorganization to avoid the overlap between profit centers, which is often found in business and industry. Figure 15.1 shows a frequent misinterpretation of profit center authority and responsibility.

- *If* a profit center is overlapped or undercut by another profit center (even worse by several),
- *then* the accountability of the overlapping centers' managers becomes at best opaque and at worst nonexistent.

When this happens, and it often does, the result is worse than having no profit center organization in the first place. Moreover, in creating a

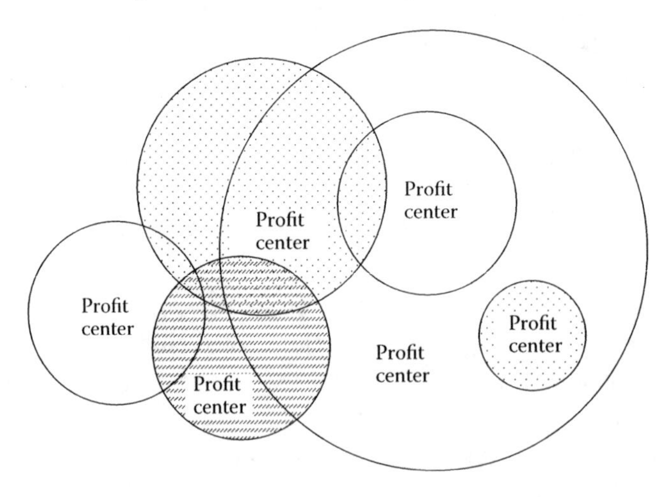


**FIGURE 15.1**
The overlap between profit centers as well as among profit centers and cost centers destroys their accountability.

profit center structure the best goal is that of including all activities falling within it *as if* each of these profit centers was a stand-alone enterprise. Close consideration should be given to the task of making them both effective and economical.

As with any other organizational mission, the focus must be on the authority and responsibility of each manager, and how well he or she performs the job to which he or she has been assigned. The system must provide him with information that is analyzed at a tempo allowing him to make the right decision at the right time.

A different way of making this statement is that at each profit center organization and structure must enhance *personal accountability*, by making feasible measurements and metrics of performance supported by cost accounting procedures. Accounting per activity should compare the *cost* and *revenue* involved by

- employee,
- product,
- production facility,
- distribution channel, and
- client account

As we will see in subsequent sections, costs matter greatly. The proper cost study should start with a thorough organizational review because quite often poor organization leads to inherent inefficiency and this is found in the origin of high costs. Does the current structure reflect the profit-and-loss guidelines of *our* firm? Is it compatible with the lines of authority and responsibility so that P&L evaluation is documented? Furthermore:

- Is it fully updated, as structural changes take place and product lines evolve?
- Is the profit center structure effectively using standard costs?* If not, what's the cost basis?
- How often plan versus actual comparisons take place? Is corrective action taken immediately thereafter?

A profit center–based individualization of cost and revenue, as well as their tracking, is not feasible when the organization allows overlaps

---

* Section 15.3.

between profit centers—whether this happens from a managerial or from a technical viewpoint. Neither is it possible to measure all cost and revenue associated to a given profit center unit when fixed, semivariable, and variable costs (Section 15.3) have not been sorted out.

Profit center reporting must be focused and timely to permit *corrective action*, in case of deviations from financial plan (the profit center budget). Costs are nearly impossible to control in case the details do not include all costs' chapters, their origin, and destination—so that the appropriate debits/credits can be made.

The analysis of historical cost data associated to each profit center, and comparison to those of other similar profit centers, can be revealing. An analytical examination yields a valuable picture of *trends* in the growth of expenses, but this requires experienced people able of providing factual and documented answers to queries.

- How much cost me the products I sold, all expense items included?
- Whose overhead am I supporting? How much is this promoting *my* deliverables? Shall I accept such overhead?

For any practical purpose, this is *performance measurement,* which allows the manager of a profit center to know himself and his operations. "If you know the enemy and know yourself, you need not fear the outcome of a hundred battles," said Sun Tzu (circa 500 BC; see Chapter 1) and he further said, "If you know yourself but not the enemy, for every victory gained you will also suffer a defeat" and "If you know neither the enemy nor yourself, you will succumb in every battle."

Performance measurement tells us about ourselves, and our achievements, as well as about our company and its achievements. It also brings in perspective our shortcomings. Once we know our shortcomings we can correct them through greater efficiency supported through training, organization, technology, and tough management decisions.

In its high time Toyota could produce twice as many cars per employee as any U.S. automaker, through better organization and management control (which subsequently waned). It used technology to capitalize on one of the auto industry's most important trends: consumer demand for greater product variety. But then second raters took over, the goals got diluted, and the company nose-dived. Quite similar situations prevail in practically all industries.

## 15.2 COST CONTROL

The way Henry Ford had it, "Small savings made in big number represent a large sum of money." "Take care of the smallest penny," Andrew Carnegie, the king of steel, used to say, "and the pennies will take care of themselves." For his part, John Rockefeller applied two principles in his way to success:

1. Drastically competitive pricing against his opponents in the market
2. Controlling the whole chain of production, which further allowed to compress costs (Carnegie had done the same)

The principle of low cost is applicable through the whole chain of industry; from production, to distribution and marketing, says Ingvar Kamprad, who built Ikea. And he adds, "Diminishing the quality to cut costs is not a solution; it's a trap." When we talk of cost control we mean that this must be done while sustaining high quality. Wise words. If the industrialists are not aware of this principle, their customers are.

The fact that customers have become cost conscious is good news for dynamic, well-managed enterprises. Quoting from Sam Walton, one of the most successful businessmen of the post–World War II years, "It goes back to what I said about learning to value a dollar. It is great to have the money to fall back on. But it is probably time to move on—simply because you lose touch with what your mind is supposed to be concentrating on: Serving the customer."*

The importance of cost control is no news to the reader. We have talked about it in Chapter 13. Commenting on how and why companies fail Walton suggested, "Their customers were the ones who shut them down. They voted with their feet." They did so because at a time when low cost is king these companies have been uninterested or unable to cut their costs and their prices. Instead, they tried to come up with some "brilliant" new engineering feat in the hope that it would fool the market.

Contrary to this attitude, well-governed companies go through heaps of market research before developing a new product, and marketing is asked to come up with a price target, which engineering and production should use at sign post. That is the winning formula. In a winning design,

---

* Sam Walton, *Made in America: My Story*. Bantam, New York, 1993.

engineering and manufacturing scrap expensive features while adding user-friendly but cost/effective elements.

A ground rule for success in the global economy is to set very aggressive cost goals and still project attractive products customers yearn to buy. iPod is an example. The No. 2 rule is that shipments must soar without major hiring; and the No. 3, that when a product or marketing approach is not paying off, it should be killed fast.

Cost-cutting aimed to increase competitiveness does not need to be confrontational. But it takes plenty of preparation to overcome a long strike. After Japan's Bridgestone bought Firestone Tire & Rubber in 1988, labor relations were downright friendly. But the coziness vanished after Firestone posted a total of $1 billion in losses by 1991. This brought a new cost-cutting strategy and a new management to head Bridgestone/Firestone.

To develop a war chest inventory, Firestone bought 500,000 tires from rival Cooper Tire & Rubber, and Bridgestone in Japan agreed to supply another 3 million tires. Thereafter, management demanded that the United Rubber Workers (URW) labor union breaks with the contract it ratified with Goodyear Tire & Rubber, which called for a 15-percent increase over 3 years.

In its place, Bridgestone asked for 12-hour shifts, pay hikes tied to productivity goals, health co-payments, and lower new-hire pay—all to overcome big cost disadvantages against competition. And though the union went on strike, Bridgestone's stockpile and production from its other factories saw to it that deliveries were unaffected by the strike. This allowed it to keep most major customers supplied, almost *as if* the strike never existed.

Then, on January 4, 1995 Bridgestone announced that it had hired 2000 permanent replacements. The move struck a nerve in the Clinton Administration, which supported a law to limit employers' ability to replace strikers. Then Labor Secretary Robert Reich, who had intervened in six major strikes, tried to get the two sides talking again. Instead of quietly agreeing, Bridgestone rebuffed Reich, saying its president was unavailable and offering its labor chief instead. The Clinton administration found itself boxed in a corner.

Bridgestone was not alone in trying to get the upper hand over costs. Cost-cutting by IBM also makes interesting reading. Because of tough cost-cutting procedures that hit every function, from order-taking and delivery to warranty management, IBM was able to significantly reduce its costs. The overhaul was prompted by a crisis during the early 1990 as the company's market share in Europe slid from more than 15 percent to barely 12 percent.

In fact, IBM got so close to the edge it really shook up management and William E. McCracken, the U.S. distribution chief, was dispatched in early 1991 to "fix" Europe.* He brought all PC functions under a single manager, giving him full reign over sales, marketing, distribution, finance, and manufacturing with strict orders that costs must be significantly trimmed.

Inefficiency prevailed for instance in Greenock, Scotland, despite Scotland's lost-cost, highly skilled labor force. Greenock used to cobble together PC motherboards for all its models on a single line, with the result that switching from one model to another could idle production for up to 8 hours.

Effective cost-cutting required restructuring as well as keeping an eye on quality of produce. After splitting the line into six mini lines, one for each product family, total manufacturing and testing time has been slashed from five days to less than 8 hours. At the same time quality rose by 10 percent.†

A similar improvement was done in distribution where costs were reduced from $200 to $45 per computer. Other major strides came from paring inventory to the bone. By shipping finished goods daily to customers direct from Greenock, therefore bypassing 14 country warehouses, IBM further cut its cost.

Thanks to new flexible manufacturing that can quickly shift from one model to another, a policy was established that Greenock builds only what is in demand. That kept critical parts from sitting around in unsold systems. As a result, the inventory of completed systems plummeted by 90 percent, to 30,000, while sales increased by 42 percent over the same time period.

These are real-life examples the reader will be well-advised to keep in mind, because the same problems have the nasty habit of popping up time and again. In Bridgestone/Firestone and IBM's action, the objective has been cost/effectiveness—the policy of identifying the alternatives that yield the greatest effectiveness in relation to the cost incurred.

---

* *Business Week*, December 13, 1993.

† Idem.

There is plenty of rationale behind this objective. Each program, each project, each facility uses up resources that could otherwise be put to some more useful purpose. Hence, programs, projects, and facilities should be periodically scrutinized with the aim to improve quality, reduce the time lag, and cut cost. Let's always keep in mind that *cost* represents effectiveness foregone elsewhere.

Companies respecting themselves and their clients find the metrics enabling them to tune their cost/effectiveness. The careful reader will recall the practical example on capitalization of IBM's products. Big Blue defines it as follows:

Direct Labor (DL) + Direct Materials (DM)

Management does whatever it takes to keep it at 15 percent of list price. To make up the cost of goods sold, to this basic DL+DM cost should be added R+D expenditures, cost of productivity enhancements, manufacturing engineering, new tooling, depreciation (factories, machines), and manufacturing overhead. Also the company's profit at technological side.

Correspondingly, the cost of sales will include the costs of branch office network, sales engineers, systems engineer, transport, warehousing, installation at customer premises, test period till "ready for use," other nonchargeable expenses, sales overhead. Also profitability at sales side. (Costing as a culture is discussed in Section 15.4, and standard costs in Section 15.5.)

## 15.3 FIXED, SEMIVARIABLE, AND VARIABLE COSTS

Not all costs incurred by a company have the same permanence in its balance sheet. Some closely related to its level of production (products or services) are variable; others, particularly those connected to land, factories, and equipment are fixed. There is as well a third category fitting between them, the semivariable.

As this brief reference suggests, our cost evaluation will be accurate only when we establish the profile of each source of cost, and are able to control its variation. Managements who don't know the costs for each product and service they offer, are taking unwarranted risks with their market positioning and profitability. In this section we will examine the following three classes:

1. *Fixed costs.* These do not change as the rate of output varies. Examples are bought equipment and (up to a point) overhead.
2. *Semivariable costs*, which do change with the rate of output but by big steps, then hold steady over a range of output. An example is adding new machinery to increase production capacity.
3. *Variable costs.* As their name implies, they vary with changes in the rate of output, and therefore they should be subject to steady watch.

Typically, variable costs are operating expenditures bearing almost a proportionate relation to the rate of output. For instance, the amount of input required for production: Director labor (DL) and direct material (DM). Notice that director labor represents expenditures employed in product fabrication—not for other purposes.

Variable costs are incurred because the company decides to manufacture a certain product, provide a given service, or execute a contracted job. Once the decision has been made to produce goods or take on a contract, the firm is obligated to engage in certain costs that are controllable only through the production volume and, most specifically, the efficient use of

- manpower, and
- material

Reports on variable costs connected with direct labor and direct materials should be drawn up quickly to permit taking rapid action to keep actual production costs in line with planned costs. By contrast, in poorly managed companies cost reports are usually delayed, appear long after the actual events have taken place, or are altogether nonexistent. Or, alternatively, they are used more to control foremen and other persons than to control the actual figures themselves.

Fixed costs, also known as period costs, are best illustrated by such items as land, building, machinery, office equipment, computers—and by extension property taxes, insurance, rental, depreciation charges, and other amortizations of prepaid expenses. Overhead is also a fixed cost (Chapter 16); though sometimes it may be semivariable, it is never really a variable cost.

Up to a point, fixed cost outlays tend to be the same whatever the rate of output may be. A very expensive machinery that has been bought for a project that did not materialize must continue being depreciated whether it works full time, half time, or not at all. Changes in the rate of production have little or no effect upon the amount of the fixed costs outlay.

Expenditures in this category are referred to as fixed expenditures, not because they are permanently inflexible in amount but for the reason that, other things being equal, a change in the rate of operations of the firm will not result in a change in the amount of the outlay.

- Normal fixed costs are depreciation of all property, plant, and equipment, as well as taxes on property, plant, and equipment.
- In extremis, some fixed costs will be expenditures that are chargeable against current profits, but cannot be identified with any given job.

Fixed costs can however be reduced. For instance by selling idle machinery at bargain prices. But management must make a decision to do so, which is another way of looking at the difference between fixed and variable costs.

Short of such deliberate management decision, accountants take the position that within normal levels of production the total money of period costs may remain reasonably constant. A controversial practice, to my judgment is that fixed costs, which include top level overhead items such as salaries, are paid to officers of the company.

Telephone expenses and utilities are often taken as part of fixed costs because they assimilate with expenses of an administrative nature. Other fixed costs are interest expense and different dues. Also, insurance costs of all types, except on payrolls. However, the way to account for some of these fixed cost items is not cast in stone:

- Nonrecurring costs are typically considered fixed costs, even if these so-called nonrecurring items can recur.
- Although volume does not necessarily affect the wear and tear of equipment, management decision can make repairs and maintenance of a fixed or variable cost.

The approximations accountants make in this and similar cases rests on the fact that total money spent on aforementioned costs is fairly constant and usually minor in comparison with total costs. Part of the uncertainty in cost classifications comes from the fact that the principles of industrial accounting have been extended in other domains, for instance sales.

Fixed, semivariable, and variable costs are plot in a *breakeven chart* against income. Care should be taken not only of production costs but also of sales/marketing expenses. Costs associated to the sales offices network, utility bills, cars used by salesmen, and support services are examples of fixed costs. The salesmen salaries and commissions are variable costs.

The breakeven point should be looked at as an interesting case of opportunity analysis. A sound approach to establish a flexible sales budget based upon an examination of expenses to determine how individual sales costs vary. A chart can depict the relationship of sales and expenses to show at what volume revenues cover expenses when marketing/sales costs are accounted for.

The advantage of a breakeven chart is that it shows in no unclear terms the level of revenues and of expenses for each volume of sales. It can also be expressed in terms of units of goods in percentage of plant utilized, or in other quantifiable terms. This analysis enhances the ability to establish at what point—for a given plant, sales network or other operational situation—the company can expect to start making profits.

People with experience in using breakeven charts consider them to be an excellent analytical tool. The reader should, however, notice that in the real world there exist certain reasons, which see to it that the relationship between variable and fixed costs is not so nearly perfect.

- The effects of prices, productivity, technological change, and seasonal influences may distort the linear pattern.
- Only in a general sense items of direct-type expenditure exhibit the tendency to vary with the rate of output.

We must as well account for semivariable costs which, as already mentioned, fit between the fixed and variable costs. These expenditures exhibit a combination of the two patterns already described, and vary according to a change of threshold in the rate of output—but they tend to remain stable under that threshold.

Within certain limits, semivariable costs may be almost zero. But *if* production is increased beyond a critical point, *then* expenditures, for instance for machinery, will increase abruptly to a higher level due to the rental of more space, acquisition of more equipment, and similar reasons.

Though these examples have been taken from the manufacturing industry, the underlying principles are widely valid. More or less, the same classes of cost behavior exists in the service sectors of the economy; for instance in banking. Through analogical reasoning, variable costs can be related to the number of client relationships being handled, the number of accounts being made, and other criteria.

## 15.4 COSTING CULTURE AND PROFITABILITY*

*Costing* is a permanent process; a mindset; an organizational culture. Costs are a function of management quality, methodology and technology we employ, as we steadily weed out manual operations. The measurement system that we develop, and its metrics, should be maintained and dynamically modified. A full costing solution includes

- the overall costing structure,
- development, or choice, of costing methods and standards,
- the right costing (and pricing of products and services),
- a profit and loss–oriented budget (Chapter 14), and
- documented profitability evaluation

In regard to the third of the above bullets a costing system should involve all products individually, transactions relative to these products, whole product lines, expenses associated to customer relationships, and more. Let's take transactions regarding financial instruments as an example.

Most often in banking, *transaction costs* are seen as being only the total costs—particularly the personnel costs associated to a transaction. This is wrong because there are as well technology costs associated to transactions as well as overhead, real estate, utilities, and other expenses.

Even with this simplistic approach however, transaction costs are not trivial. At the classical teller level the cost of a customer operation is in excess of $10 per transaction. One of the banks participating in this study had 100,000,000 transactions per year, adding up to staggering costs. More complex transactions have a higher price tag associated to them. A British bank calculated the single Forex transaction costs to be £20 ($30).

* See also Section 16.3.

The right costing of a transaction should include direct labor, equipment, utilities, and overhead. Automation is not always successful in swamping human costs. Even if we weed paper out of the system (which is far from being a general case), human costs stay in. A true cost control effort should focus on taking human costs out of the system.

Precisely because transaction expenditures are nontrivial the right costing is very important. From chief executives to managers, professionals, and clerical workers everybody must change his or her concept about costs, adopting, and implementing an efficient costing method along the model used for several decades by the manufacturing industry.

Globalization has brought many challenges connected to the notion of cost. The way a news item in early January 2010 had it, a small handworkers shop in Egypt went out of business because it could not anymore compete with basic farming tools, such as plows, imported from China.

Nasser al-Hamoud, a blacksmith, worked from inside an old mud-brick building on a twisting narrow street in Al Qasr, a village at the northern end of Egypt's Dakhla Oasis. Hamoud was smithing there for a quarter century having learned the trade from his father, but he said that business is uncertain these days. Farm tools were his bread and butter, but cheap imports from China have flooded the market. (He scornfully compared a thin Chinese-made shovel blade and his weightier but more costly hand-forged version.)*

To survive the Chinese low cost competition, Hamoud began a sideline in selling large ornamental nails and rustic jewelry to tourists, who had started to visit in larger numbers. He did so out of necessity, though he had no idea if this was a blip or a trend. That is a bridge too far for the Chinese who, after having sweeped clean jobs and companies in America and Europe, now do just the same with the manpower of less developed countries. But it is also an evidence on how much cost conscious not only industries but also artisans must get in order to survive.

It is hard to find a story better able to dramatize how much costing must become a culture, in order to survive in an age of globalization. Even the suckers are wiped out.

Anecdotal evidence suggests that Joseph Kennedy, who made a good part of his large fortune by selling at the stock market high in 1929 just before the crash, got his tip from a shoe shine boy in Chicago. As the boy was polishing Kennedy's shoes, he asked, "Eh, Mister, you want a tip for

* Peter Kenyon, *Eco-Tourism Holds Promise, Peril for Egyptian Oasis*, NPR, January 6, 2010.

the stock market?" Kennedy thought if stock market fever reached shoe-shine boys, who are going to be the next suckers. Thinking by analogy

- *if* Egyptian handworkers have to close shops because of Chinese lowest cost wares,
- *then* who will be the next suckers? Quite probable the Chinese workers themselves

The Chinese Communist/Capitalist party bosses should think of that, and the same advice applies to all boards of directors and CEOs in the East and in the West. There is simply no way to be lenient about costs and survive. But not all CEOs and, surely enough, not all governments appreciate the importance of careful scrutiny of costs. Here is a startling example.

On the excuse of protection against terrorist attacks, after 9/11—which was, indeed, a necessity—George W. Bush demolished the efficient U.S. Consular service and privatized many of its functions (which has been an aberration). In the aftermath, America's embassies became fortresses and the procedures much more stringent but necessarily more effective.

Here is how an article in *The Economist* described this high water mark of bureaucracy: "Applications must be submitted online, accompanied by a non-refundable $131, paid electronically. In return, the applicant receives a confirmation e-mail, which includes a barcode... (plenty of) absurd questions are asked but the answers are never verified."*

It needs no explaining that bureaucracy is the deadly enemy, at the same time, of efficiency and of the costing culture, with plenty of added costs hitting both the applicant and the U.S. taxpayer. Bureaucracy adds a great lot to other practical existing difficulties with costing:

- Lack of top management appreciation and backing
- Muddled ways in the handling of overhead
- Lack of collaboration by operating departments, some of which thrive on inefficiency

In the background of all that lies the fact that both leadership and control action expenditures skyrocket and profitability is put in question. Whether we talk of a government or of a company, the cost of spoilage and of unnecessary expenses comes out of the same purse where reside the profits.

* *The Economist*, February 16, 2008.

To enhance our company's profitability our goal must be to measure the cost and price attached to each product and service; as well as individual customers and total customer relationships. To do so, we need two components: statistics and projections. Both should focus on the business generated by

- customer,
- channel, and
- product or service

This requires an effective costing scheme based on standard costs (Section 15.5). Reaching this aim also presupposes a good method for analyzing the origin of costs and of their variation. In a costing study that I did some years ago for a financial institution, we established the following parameters for a system able to track bank profitability:

- Setting and observing profitability standards
- Knowing our costs and having a policy to control them
- Putting emphasis on the 20 percent, high net worth clients who produce more than 50 percent of profits
- Automating as far as possible the handling of the other 80 percent

Part of the overall strategy has been strengthening relationship management through investment advice, which reintroduces direct labor costs but also produces a higher return on investment. This has been calibrated through a system of performance information that also provided hindsight on "do's and don't's" along organizational and product lines.

The cumulative effect has been to highlight the role of *individual responsibilities*, while using technology to aid management in planning, measuring, and controlling. The information flowing in and out of the overall cost control and profitability tracking aggregate is consistent and relevant to the structure followed up by *responsibility centers.*

The net effect has been to raise the value of operating data available for operations and for control purposes. Profit center reporting has been put in a position enabling evaluation of organizational profitability, with set targets and alarms triggered by exception. With some minor exceptions of special operations, the overall financial reporting structure uses standard costing.

## 15.5 STANDARD COSTS

Standard costs are standardized metrics of labor-oriented expenditures associated to a wide variety of work elements pertinent to the company's operations. As such, they reflect themselves in products and processes, particularly in connection with activities associated to

- production, and
- distribution

Down to basics, standard costs are predetermined costs associated to standard times and tabulated. These standard times are conforming to the specifications of the product, but must be recalculated when the production process changes due to automation, alteration in product design, or any other event affecting direct labor figures. This is done through analysis of causes of failures in cost control. From a managerial viewpoint, the object is that of

- having a standard against which is compared the actual cost, and
- alerting on deviations from observing this standard, when it is still time to bend the cost curve

Even on the same job, standard times can vary as a function of the degree of automation of the job—but also for the same level of technology because of differences in training each operator underwent to efficiently use the technology put at his disposal. Another reason of variation is the quality of management supervision.

Standard costs are *unit cost* and their estimates should be realistic rather than optimistic. The degree of accuracy and number of activities covered by a standard cost study depend on what exactly is wanted, and on the measurement technology being chosen. The better known measurement techniques are as follows:

1. *Work sampling*, a form of random sampling of work activities practiced in offices and factories for several decades
2. Tables that provide *preestablished elemental times*—like method time measurement (MTM)
3. The time-honored *motion and time study* with stop-watch, *Therblig* analysis for greater precision

One financial institution to which I was consultant established the following work sampling element classes for cashier/teller operations: passbook-in, passbook-out; cash received, credit of an account; cash paid, debit of an account; other payments; foreign currency, sale; foreign currency, purchase; cash check, own branch office; cash check, other banks; counting cash on hand; checking cashier's tickets. There were as well reserves made for less well-identified operations and delays.

These are typical retail banking operations and in a real-time environment most of the work elements in the aforementioned list are traceable by computer. Expert systems carefully analyzed, for management reporting plan versus actual cost figure for each work element by cashier position—summing up all the way to products, processes, and bank workers. Other expert systems evaluated each product against its standard cost, flashing out deviations. They also differentiated the corresponding income from charges related to teller operations, between

- real price, and
- virtual price

The real price is always market-based; the jargon "virtual price" identified the cost of the product to the bank. This has been used for comparative reasons between real and virtual price—as well as for tracking virtual prices between branch offices. As experience was gained, 50 banking products have been selected for the second phase of the standard cost study, with the aim to clearly sort out where exactly the employee's time goes. These fell into three main product lines:

- *Treasury*, namely operations characterized by direct treasury involvement: Accounts, portfolio (commercial, financial), and more
- *Signatures*. This is particularly important for obligations entered into by the bank in the form of guarantees; they involve risk but no direct treasury transaction
- *Services*. This class included what the bank characterized as "risk commitment." For instance, payment of a check, the handling of a currency change operation, documentary foreign trade activities

Top management's thinking in connection to the standard cost implementation is best reflected through the CEO's statement, "Cost accounting should hit where it is essential—and essential means where costs are high.

Or can run out of control." But though the most *important* were targeted, the system was built bottom-up. Management started the project with the question,

- what constitutes an elementary banking operation?

The analysts answered, "the handling of a check," and they subsequently chose a unit of measurement for the other handling of a single check. Subsequently, a group of somewhat more complex operations were costed as a multiple of handling a single check.

Let me repeat this reference. To gain standard costing experience in a first time before the more elaborate standard costing study was undertaken, several other banking operations have been established as multiples of this elementary *unit*. Commercial paper handling, for instance, was taken as equal to 10 units.

After experience was gained over a period of less than three years, a small group of analysts and information scientists working together developed a standard costing system for the 80 percent of all jobs at

- headquarters,
- regional offices, and
- the network of smaller branches

Another study in the same credit institution compared the results of work sampling and self-rating in the wealth management department. As shown in Table 15.1, these results have been nonconvergent. The differences

**TABLE 15.1**

Comparative Study of Results from Work Sampling and Self-Rating

| | Work Sampling Observations | Self-Rating Questionnaire |
|---|---|---|
| Meetings | 34% | 28% |
| Coordination and Supervision | 24% | 7% |
| Report Preparation | 8% | 18% |
| Phone | 7% | 5% |
| Information Gathering | 6% | 21% |
| Traveling | 4% | 4% |
| Miscellaneous | 17% | 17% |

are due to the fact that managers think of themselves as planners with their supervising role being rather minor. In reality, however, the opposite is true.

Four principal factors determine the nature and detail of techniques to be used in a standard cost study. The first is the extensiveness of the job, measured through the number of man hours per day or per year used in a given type of work. The second is the *anticipated life* of the job, which helps determine the detail we should be after and hence the investment to be done. The third is *labor share* in a particular operation, such as

- ratio of handling time to machine time, and
- required special qualifications of the employee

Another criterion is investments projected to be made in machines, tools, and equipment to be applied to this job. *If* new investments are contemplated, *then* precision is not so important, because after new equipment is acquired and installed the standard costing study must be renewed.

As this and plenty of other experiences are documenting a properly implemented standard costs study reflects fairly accurately the true cost of providing services to bank customers, and permits to price products for a fair return on capital. It also provides a common denominator; a *yardstick* by which profitability, or lack of it, can be measured.

Standard costs can as well be used in management accounting for indicating *cost trends* quickly, by making actual costs more useful as they show when and where they are out of line. As such, they provide the basis for better control over expenditures, focusing attention on the production and distribution cycle.

A further advantage of standard costs is that they see to it that control rests on a documented, analytical ground. The example in Table 15.2

**TABLE 15.2**

Control of Personnel through Standard Costs

| Salary Class | Actual | Standard | Variance |
|---|---|---|---|
| A | 1 | 1 | – |
| B | 7 | 4 | +3 |
| C | 7 | 6 | +1 |
| D | 2 | 4 | −2 |
| **Total** | **17** | **15** | **+2** |

comes from the same financial institution mentioned in the beginning of this chapter, and concerns the redimensioning of the loans department in one of the bank's regional branch offices.

As it can be appreciated from the *variance* in this column, the department had two employees more than it needed. Most importantly, highly paid personnel (Class B) were used to do the work of much lower paid personnel (Class D) whose number was insufficient.

This and other applications in the realm of management control have been integrated into the bank's *responsibility accounting* system. Standard costs identify for every manager individual human resources responsibilities for *control* purposes. Profit center reporting is also using standard cost data, converting them into a "contribution margin" approach through basic measurements.

At top management level, the board and CEO have been receiving product line reports that include standard costing and permit to compare the relative value of products and services. Expense chapters allocated to profit centers and specific products have also been supported through standard costing. To the contrary, overhead costs were not allocated, but isolated at the level of the organization where they incurred—and brought to senior management attention at that level, with the objective to swamp overhead (Chapter 16).

# 16

# *Financial Planning and Control*

## 16.1 LONGER-RANGE FINANCIAL PLANNING

Chapter 14 treated the broader perspective of financial management, concentrated on the one-year financial plan: the budget, as well as on cash flow. Chapter 15 addressed itself to profit centers, cost control, and standard costing. The object of this chapter is to present to the reader the longer-term view of financial planning into which integrate the choice of debt versus equity, profit planning premises, management accounting, and financial reporting.

The purpose of a longer-range financial plan is to match the further-out objectives of an enterprise with its financial staying power, and based on this, to calibrate top management policies and decisions. Precisely for this reason, integral part of longer-range financial planning is *profit analysis*. The term stands for continuing review of costs and profit margins so as to ensure the realization of objectives connected to

- steady cash flow, and
- a sustainable profitability

As the careful reader will recall, *cost control* has been singled out as an important constituent of sound financial management. Reaching cost objectives calls for a top management policy backed up by the steady accumulation, analysis, interpretation, and reporting of detailed plan versus actual unit cost data, the identification of deviations, and appropriate corrective action.

Chapter 15 also brought to the reader's attention that closely connected to cost control are *manpower control* and *overhead control*. Both require regular analysis, interpretation, and reporting of the utilization of labor at all echelons as well as a policy of productivity improvements—which is connected not only to standards but also (if not mainly) to the steady upgrading of skills and know-how.

A high level of *equipment utilization* through optimal capital expenditure decisions, is another goal of longer-range financial planning. It requires timely reporting on scheduling and use of company equipment, projection on forthcoming needs, and a steady updated equipment replacement policy. Equipment purchases at the spur of the moment should be strictly avoided.

Inventory policy is without any doubt integral part of longer-range financial planning. Its alter ego is a policy of *inventory control.* Its implementation requires the development of inventory targets, timely reporting of their status, maintenance of online databases with accurate records, and mathematical models that enable just-in-time inventory management.

Good governance is based on the assurance that all of these basic elements are incorporated into the financial planning and control system. They must be adapted to the company's specific operational requirement and refined in a way permitting to set and re-evaluate management policies concerning longer-range assets and liabilities management (ALM).

The best approach is to match ALM with risk management using knowledge artifacts, which provide the linkage, as well as the input to an exception reporting procedure. For instance, in a bank's assets and liabilities management swap income and expense are reported in interest revenue or interest expense, applicable to the related assets or liabilities. Yield-related payments or receipts associated with swap contracts are accrued over these terms of the contracts. But risks associated to these transactions are not apparent—and this is a bad management practice.

True enough, if not closely matched with offsetting swaps, such contracts are marked to market value, with changes in market value recorded in trading account profits and commissions. Marking to market, however, is an inexact art, and the law of the mind if often manipulated by politicians.

(For instance, during the 2008 deep economic crisis the arms of Financial Accounting Standards Board [FASB] were twisted by the government to cancel marking to market reporting. The irony of this is that the U.S. government considered marking to market untouchable as long as it contributed to fat profits and bonuses for the bankers.)

A parallel ALM/risk management system to be developed for management accounting only, will consider both the accruals and marking to market procedures—taking every time the lower of the two values and matching losses with profits (if any). The underlying concept is not too different from legally required, that

- fees on interest rate swaps are deferred, and
- they are amortized over the lives of the contracts*

Sound governance also brings in perspective the need that the assets over liabilities ratio should be greater than 1. The importance of A/L can hardly be overemphasized. While the exact metrics to watch vary by industrial sector, there are as well factors of much more general applicability to which all companies need to pay steady attention. In alphabetic order:

- *Assets.* Total assets, backed by details, should be available to management not only as of end of calendar or fiscal year but as well monthly and immediately on request.†
- *Capital.* Information on the company's capital plus extraordinary reserves should be interactively available to all authorized persons.
- *Cash Flow.* It should be presented discounted to net present value using the going rate for the cost of capital plus pertinent risk factors (see also Chapter 14).
- *Debt.* All types of debt exposure assumed by the company (loans, bonds, and so on) should be available not just as of end of calendar or fiscal year, but also monthly and on request.‡
- *Liquidity.* The important liquidity figure for management accounting is money in the bank plus liquid items in the portfolio: bonds, equities, liquid derivative contracts, and other investments that can be quickly turned into cash (after a haircut).
- *Market-to-book.* Ratio of market value of common stock to book value of common stock.
- *Net income to net interest income.* Ratio of net income (for calendar or fiscal year, adjusted for reporting differences) to net interest revenue. (This metric is useful in the banking industry).

---

* The exact nature of this and similar operations is defined by accounting standards and changes over time.

† A good proxy of assets for snapshot evaluations of the company's staying power is its market capitalization.

‡ Total face value of all outstanding loans, net of reserves for loan losses; idem for bonds.

- *Off-balance sheet risk*. Exposure in derivatives, in notional principal amount converted into loans equivalent.*
- *Return on assets* (ROA). Net income divided by average assets.
- *Return on equity* (ROE). Net income divided by average equity.
- *Return on revenue* (ROR). It is calculated as net income divided by revenue.

Classically, companies watch very carefully year-on-year the change in ROA and ROE. Better results are obtained when the year is characterized by strong net income growth, cost control, and steadily improved risk management. Return on revenue is not very popular metrics, but it provides eye-opening figures. ROR measures a company's profitability and also provides interesting information on the incumbent management's success or failure.†

## 16.2 DEBT VERSUS EQUITY

"Debts are assets," said Henry Kaiser, the famous American industrialist of the 1930s and World War II years. "It is through debts that you get the money to meet payrolls and buy the plants and equipment. Then you finish the job and pay off your debts and everybody is happy."‡ Notice that Kaiser emphasized "*pay off your debts*." This is not what happens today as debt has become a mountain higher than Everest—at family, corporate, and government level.

Kaiser looked upon business as a game with all the excitement of both winning and losing. From his early days as constructor of the Hoover Dam, he knew there will be some jobs where he will make money and other jobs where he might lose his shirt. But he was an entrepreneur who knew how to manage risk. The challenge was always to come up with enough good-paying jobs to help his weathering the bad times. Therefore he never let debt get out of control, but remained in charge

Exactly the same principle applies with investments. "In a low, one should wait for the market to turn" said Amadeo Giannini, the great banker. Giannini strongly advised his clients, friends, and employees

* D. N. Chorafas, *An Introduction to Derivative Financial Instruments*, McGraw-Hill, New York, 2008.

† www.totaltele.com, October 2007.

‡ Albert P. Heiner, *Henry J. Kaiser: Western Colossus*, Halo Books, San Francisco, 1991.

never to borrow money to play in the stock market. If they were not in debt, he told them, they can sit on their shrinking assets and wait. They could even buy equities at bargain prices. The market will turn.

Being in charge of debt made all the difference with Kaiser, Giannini, and other great leaders of industry. This has probably escaped the attention of theorists (Miller, Modigliani, et al.) when they preached that companies should prefer debt over equity. The criteria backing up this "love your debt" theory are at best asymmetric. Its developers, who are economists *not* businessmen, suggest that the choice between debt and equity should depend on three factors:

1. The rates at which debt and equity are taxed
2. The potential costs of bankruptcy
3. Conflicts of trust between managers, shareholders, and creditors

True enough, tax laws perversely promote debt over equity, but not only the potential costs of bankruptcy can be very high—all the way to oblivion—while the fact is that a company's goal is to perform, not to go bankrupt. Look at Bear Stearns, Lehman Brothers as well as Citigroup, Bank of America, Fanny Mae, Freddie Mac, and AIG; the latter three "saved" at the 12th hour through massive injection of taxpayer money.*

As for conflicts of trust between managers, shareholders, and creditors they become most real when the market turns sour. Then, every stakeholder wants to get back his money and run.

- Debt becomes a poison.
- Only equity provides a sense of stability.

Debt exacerbates the so-called *agency costs*, that is the conflict of trust between managers, creditors, and shareholders. These can have a significant effect on the course events will take. Managers often know far more about their firm's prospects than lenders and investors who put up the money. For this reason both lenders and investors are constantly looking for more precise information, but creative accounting turns the balance sheet and income statement on its head.

* Still, in spite of massive public money injection Fanny Mae and Freddie Mac were delisted from the New York Stock Exchange on June 16, 2010.

Factors such as amounts of debt, taxes, share prices, as well as the lack of reliable statistics and financial information play a major role in how fast a cash starved company goes against the wall. Failures induced by excessive debt in the years the Modigliani–Miller theory was aired as the new wonder, saw to it that during the last few years financial analysts and economists have increasingly come to look in a more rational way on choices between debt and equity.

This change of heart toward debt has been strengthened by the huge risk of solutions based on leverage, therefore debt, and the accelerated amount of failures in the aftermath of the 2007–2010 deep economic and banking crisis.

Classically companies take on debt in order to benefit from the fact that interest payments are tax-deductible (in some countries). They also issue debt in the form of bonds to profit from discrepancies in interest rates and conditions between a bank loan and a debt instrument—as the capital market has its own credit standards. But there is a huge difference between

- seeking financing, and
- becoming a debt addict

At least theoretically, a firm whose true value is underestimated by the stock market may do better by borrowing funds than by issuing more equity at an undervalued price. Also, many people assume that by replacing equity with debt a firm can boost shareholders' returns on the remaining shares (which is a false assumption).

These, however, are not unchallenged principles. Ironically, they were practically shattered by Dr. Franco Modigliani and Dr. Merton Miller themselves when they showed that, in the absence of other factors such as taxes and bankruptcy, the value of a firm's assets would not be affected by whether they were financed by debt or equity. (This, too, is a false statement.)

No wonder therefore that the supremacy of debt is now challenged in more than one quarter. A shrinking value of equity makes it cheaper for shareholders to own a significant part of their company's shares. It also allows shareholders to give management an equity stake cheaply, and enables a group of shareholders to monitor company management.

But above everything else, it is the instability generated by leverage that should discourage the use of debt as a way of doing business (let alone as a hope of staying in business). Already in 2001, way before black year 2007, an article in Business Week had this to say about debt: "If Verizon, with its $60 billion in debt, runs into troubles funding third-generation wireless networks, deep-pocketed Gent (then the boss of Vodaphone) would step in—and win control of the North American operations for himself."*

To assure that the company is not running itself into the ground because of overleveraging and oversight, budgets and debts should be established by independent functions—with the latter's boss reporting directly to the head of risk management, and from there to the Risk Management Committee of the board. It is not enough that the board provides financial goals for the business. There must also be control over the assumption of debt.

Last but not least because organizations are made of people, the company's managers and professionals must be trained to understand the risks of debt. No system or function, regardless of how well it is organized, can produce beyond the capacities of its individual members. Top management is therefore well advised to look at the requirements for qualified personnel and develop longer-range programs that will keep the company out of debt as a way of doing business.

## 16.3 OVERHEAD COSTS

Overhead costs can best be defined as indirect labor. Typically, they are a lousily managed business but also tough to bring under control. In practically every company overhead costs tend to escape supervision, and therefore they must be very carefully watched. My experience is that overheads pop up like the heads of the Hydra, and to control them one must

- make them highly *visible*, and
- assure that they don't spread over, or are hidden by, different profit centers

Making overhead costs visible means showing them up exactly where they occurred, then swamping them. In one of the banks I was consultant to the president, general management overhead was first compressed then compensated through the so-called general management account. This

* *Business Week*, May 21, 2001.

benefited from the difference between "active" and "passive" interest rates in the branch offices.*

At Litton Industries, overhead was controlled by limiting the available space to house bureaucrats. Tex Thornton used as headquarters a former mansion at Beverly Hills, and there was a finite number of individual rooms, hence a limited number of bureaucrats that could be hired to run the industrial empire, since each of them had to have an office.†

At Litton as everywhere else, the control of overhead has been so important because overhead costs are eating up profit figures for breakfast, and they typically account for a significant portion of product costs. *If* their allocation is not done methodically, *then* they may misalign product prices and hit profits like a hammer. The way to bet is that "The thicker the carpet at headquarters, the thinner the dividend."

It is no more a secret to the reader that high expenses leave a company behind its global competitors. Therefore, the CEO should periodically mount a tough campaign to shed employees who are riding desks rather than horses. Slimming down is the best therapy to a company that needs a turnaround.

It is of course evident that when the goal is to cut overhead costs white-collar layoffs are inevitable. Much of the company fat has been created as the great white-collar job machine rolled right through the 1980s, 1990s, and first six years of this century. But the 2007–2010 deep economic crisis brought overhead consciousness, along with a policy of trimming middle management.

In companies, which for one reason or another cannot restructure, the sprawling middle management is a nightmare, and the same is true of other superfluous jobs. Rumor has it that Electricité de France (EDF), a government power company, is 30 percent overstaffed compared with its competitors in other countries. When this happens, restructuring and reengineering become unavoidable.

Banks and airlines share the same problem with EDF. One of the biggest and pressing issues in an airline are open labor contracts with its unions. Salaries and related costs are an airline's largest single variable expense. With high overhead airlines, it accounts for 33 percent of operating costs.

---

* Typically, in a credit institution some branches have a net excess of deposits over loans, and are paying "passive" interests on the deposits. Others give more loans than the deposits they get, gaining "active" interest. The general management account intermediated between the two.

† Remembrance from a visit to Litton headquarters in Beverly Hills, in connection to a research project with the American Management Association.

Overstaffing becomes opaque when administrative costs are hidden behind profit centers. I am personally against spreading overhead, because there is no fair way of doing so. But if one wants to spread it, then he must understand how to allocate overhead costs. As a start, the activities that drive the overhead costs must be examined, one by one:

- General management
- Real estate
- Utilities
- Handling costs
- Accounting costs (billing, invoices)
- Clerical assistance

The board should appreciate that too much money spent on overhead leaves a company with a slimmer financial base from which to fund new business, and curtails its ability to be a low-cost producer and distributor of products and services. Therefore the company must be very thrifty with items such as administrative and general office staff who do not work on the production line but hide behind more visible jobs.

A thorough cost accounting discipline can reveal where a company is really piling up overhead. Detail is very important because better understanding of overhead costs opens the door to other cost reductions. "European firms have twice as many workers as their Japanese counterparts," said some years ago an official at a Japanese electronics company, "and they have not been successful in grasping the point of view of the consumer."* (Nothing changed since.)

The profit centers principle and standard costs (discussed in Chapter 15) help in identifying where unnecessary expenses grow. Their absence is behind the aforementioned argument that overhead allocations are arbitrary. The reader must as well be aware that direct costing essentially tracks the costs of material and labor, while overhead items are period cost.

- A *period cost* is a cost incurred each period that is not affected by the volume of production.
- Therefore, there is no easy way to allocate it on a representative basis such as the number of products produced.

---

* *Time*, November 12, 1990.

This fact is disregarded by companies who apply the so-called full absorption costing, which means the allocation of all overheads to products. Often a figure such as the number of direct labor hours used in a product or process is chosen as the base for such an allocation. This means that if a product requires more hours to make it will bear a greater proportion of the overhead, because overhead is spread like peanut butter on a slice of bread.

The pros say that activity based costing has been an attempt to solve this problem because it looks for links between activities rather than ready products. But this too is arbitrary. It ends up being an uneven spread of peanut butter—which is different but not really better than the even spread, even if it looks more closely at underlying activities.

Hence my advice that the best strategy is to swamp overhead costs, not to spread them. This does away with the thick carpets and other useless goodies, but it takes guts to do it. When overhead costs are singled out on their own, rather than hidden behind productive activities, it is easier to test them, trim them, and control them. Besides that profit planning becomes much more effective.

## 16.4 SOMETHING CAN GO WRONG WITH PROFIT PLANNING

Profit planning is a longer term proposition. Profits made overnight are lost overnight, because they are based on speculation. Moreover, it is wrong to believe that year after year profits will increase because thats what the management wills. Every dynamic enterprise encounters reverses, but these are overcome when there is in place a sound profit plan.

Even speculators know this. For instance, some hedge funds managers have the decency to opt out of the game before ruining every penny in the assets investors trusted them with. Undone by bad debts on old-economy stocks and on wrong foreign exchange guesses, Julian Robertson Jr. closed down his Tiger Management fund including its flagship Jaguar fund and five other entities under Tiger's wings.

At his peak in 1989, Robertson had $23 billion under management, and ran six different investment pools. But a combination of investment losses and heavy withdrawals left Tiger Management a shadow of its former self. When in 2002 the time came to opt out, in a letter Robertson told his partners he had already liquidated Tiger's portfolio and was ready to immediately return in cash up to 75 percent of some $6 billion in investments to stakeholders. This case is most relevant because

- Robertson was revered for his long-term record as a highly profitable investor, and
- he did to his credit several *coups* that benefited his clients

In the early 1990s teaming up with George Soros he had broken the British pound, but by 2002 he appeared to have succumbed to rapid change in the world financial markets. Shortly after their high water mark in managed wealth and in deliverables, the Robertson hedge funds found themselves in the wrong side of a series of big investment bets. The first was a $600 million loss in the autumn of 1998 when Russia defaulted on its debt. This has been followed by a $1 billion loss on the Japanese yen at the end of 1998. Notice that

- the bet that the yen should move south was reasonable,
- but reason has little to do with financial markets at large, and foreign exchange in particular

In 2000, in a letter to shareholders Robertson wrote, "We are in a market where reason does not prevail." Retreating from global bets, as some other hedge fund managers also did, Robertson refocused on beaten-down stocks in the old economy, like shares of U.S. Airways, Bowater, and Sealed Air. Those bets, too, proved disappointing, as the stocks continued to lag far behind the technology sector.

Because the Tiger hedge funds lost so much money in a couple of years, the remaining assets could no longer generate enough fees to cover operating expenses. Robertson's funds were down 13 percent in the first 3 months of 2000 after a 19 percent decline in 1999. As a hedge fund manager who declined to be identified was to suggest in a meeting, "It's hard to come back from being down that far."

This is true, and a proof that somewhere along the line profit planning can become inconsequential. When this happens, the better alternative is to stop the bleeding of assets. Shutting Tiger Management signaled the end of an era for Julian Robertson, his 13 hedge funds known as Tiger Cubs, and his legacy, which ran up and down Wall Street.

The lesson to retain from Tiger Management is that the day comes when the attention paid to market trends bends, partly because fatigue sets in and partly because one thinks that he is infallible. This is the time when fortunes change and one should expect the worst.

It follows that there is no alternative to a systematic approach to profit planning. This is in itself a culture whose observance requires the right methodology. Section 16.5 explains the nature of a systematic profit planning system that I helped to develop in a well-managed industrial company.

## 16.5 A PROFIT PLANNING METHODOLOGY

A systematic profit planning methodology is based on a longer-range plan, which is an action program with several financial overtones. Because profits are no random events but the result of the company's properly planned and managed activities, prerequisite to such a program are the answers to be given to four basic questions, which are also present in the making of any other longer-range plan:

1. What the company wants to be?
2. Which commitments will be required?
3. How will management face these commitments?
4. What's the return we expect from these commitments?

Most vital to an action program for profits are the goals established by the board of directors, CEO, and executive committee for four different time periods: 1, 3, 5, and 10 years. Still another basic requirement is the clear definition of the basic ingredients used in describing a profit plan. "Profits" means different things to different people. By contrast, the following factors are precise:

- *Ordinary profit*, also known as operating profit, it is equal to operating income minus operating expense.
- *Current profit*, also known as net income, it is equal to operating profits plus extraordinary profits. From the latter must be subtracted extraordinary losses and income taxes.
- *Profit from operations* is equal to net income plus fees and commissions income, after subtracting fees and commission expenses; plus other operating income, after subtracting other operating expenses; plus interest income, after subtracting interest expenses.

When we talk of longer-range profit planning we mean all three of above reference factors, their trend and distribution of their values. Mean value and three standard deviations must be steadily computed. Notice that all three reference factors are influenced by the success or failure of action programs in terms of accomplishing the company's objectives.

One of the prerequisites of a systematic longer-range profit plan is that all of basic issues affecting it are examined separately, then integrated into a comprehensive picture. The starting point is the kind of business a company is, or intends to be, over the profit planning time period.

A statement of purpose gives long range direction to the profit plan in terms of industry orientation, markets, products, and skills. It defines the broader future course now foreseen for the business. A well thought out purpose provides a sharp focus to guide all of the other elements in the plan, and sets the stage for

- the profit plan itself, and
- its critical reviews by senior management

To be helpful in profit planning, the "statement of purpose" must provide a clear definition of what the board and CEO think of their company's business—based on product, market, production, and sales methods. And because this profit plan is longer-range it must respond to the queries such as:

- In what business will this company be in in 5 and 10 years from now?
- How will this add to, or be different from, its present business in potential? In profitability?
- Is the company expecting to have greater control over markets? Over prices? How will this be achieved?

Answers can be made meaningful only by taking into account opportunities and pitfalls of the industry *our* company is in as well as the capabilities and limitations of the enterprise itself (see also the Tiger Management case study in Section 16.4). The industry, the company, and the aspirations of management must all be analyzed in determining a longer-range purpose and its impact on projected profits.

Stated in different terms, the making of any serious longer-range profit plan is indivisible from the future "position planning" effort the company makes, including measurable targets set at a specific time in the future—and, most specifically, profit targets. Some of the goals may be reachable beyond 5 years into the future; these will generally be stated in nonfinancial terms, such as share of market or product sales ranking in the industry. Even so, they will still reflect profit aims.

- What percent of market and what market rank this company plans to attain? In what product or product category?
- What sales volume such estimates represent? In which markets? At which prices and level(s) of profit margins?
- What new market areas *our* company plans to enter? When? How? In what product categories? What's the projected timing of such moves?
- What new skills in R&D, manufacturing, marketing, and finance must be obtained? How fast? At what cost?
- What increase in productivity is *our* company targeting? In which areas of operations? How this compares with competition?
- What increases in investment will be required? How much? When? For what product areas? At what ROI?

Longer-range *profit planning* should follow the company's master plan guidelines, because these chart the course to be followed in moving toward the further-out goals. They also interrelate the various efforts required to accomplish long range objectives most efficiently, providing a pattern for allocation of resource as well as selection and timing of actions. Other critical queries affecting longer-range profit planning estimate regards policies the company plans to follow in relation to

- pricing,
- make or buy decisions,
- product quality,
- brand name(s),
- distribution efficiencies,
- personnel training, and
- organizational effectiveness

It needs no explaining that the items in this list will lose much of their interest and impact without the proper consideration being given to *timing*. The timeframe over which planned changes must be fitted together in a logical order to form the overall strategic plan, is most important to every longer-range profit plan.

The same is true of management intent regarding investments the company is committing to for the next 3, 5, and 10 years; together with financial estimates of how well costs will be met by the projected cash flow (Chapter 14). Last but not least among the factors affecting a systematic longer-range profit plan is return on investment (ROI).

Even some of the better managed companies are not really sure regarding the return they will receive from their investments. Investments in IT are a classical example. Every year, companies are spending billions of dollars, pounds, and euros on IT but very few boardrooms know

- the value of hardware and software their firms acquire, or
- what a contribution IT investments are to their business successes

In spite of this absence of cost/benefit analysis and associated profitability evaluation, or even of simple understanding of IT's contribution, in America and in the EU company spending on hardware and software increases significantly year after year. It needs no explaining that this reduces by so much current and future profit figures since the money destined to low ROI comes out of the company's treasury. Everything counted, this failure can be debited to the lack of a system of personal accountability.

## 16.6 THE PLANNING, PROGRAMMING, AND BUDGETING METHOD

The planning, programming, and budgeting (PPB) method is an example of how basic personal accountability concepts can be brought together comprehensively. In terms of methodology, it draws together policies designed to control misuse of funds, and to assess work efficiency by means of effective planning. The objective is to help management make better decisions on the allocation of resources among alternative ways to attain established goals.

The name of the process derives from the fact that longer-range planning becomes tied into the budgeting-making process. The costs of programs are compared, and financial plans are drawn up for five years ahead, instead of the typical one year budget cycle. Then, for programming purposes, activities are grouped according to mainstreams identified by common purposes (and, if possible, common characteristics).

The CEO can specify in a quantitative way what he means by effectiveness or call on his assistants to use the *Delphi method,** asking a group of experts to indicate the relative merits of various alternatives and identify the parameters critical to choices among them. In output evaluation, consideration is given to competing measures, keeping constraints in perspective.

- Not all courses of action can be financed.
- It is inevitable that a large part of a budget is already irrevocably committed to previously contracted obligations.

The PPB mission is then, to devise a way for factual and documented choices comparing a finite number of alternatives, and proposing a preferred allocation of the still freely disposable funds. The Delphi method helps because it is based on polled opinions of either quantitative or qualitative estimates.

For instance, the chief financial executive might ask a panel of experts to list measures that they feel should be included in a given program. Since a "measure" is rarely, if ever, of an all-or-nothing nature, the experts also indicate the degree to which it can be executed. Through Delphi the financial executive may as well obtain estimates of effectiveness from a panel. To make communication among members of the panel possible, measurement standards are established, and evaluation is done by

- assessing the contribution of each measure, and
- estimating the percentages by which the measure in question would raise the initial conditions toward the required level of cost/effectiveness

As these references demonstrate, the essence of PPB is the development and presentation of relevant information as to the cost/benefit of a course of action. Insight derived from its usage is better evident in a wholesome effort, but PPB can as well be applied to piecemeal, fragmented, last-minute program evaluations. In all cases, it takes its inputs from classical financial tools like cost accounting and performance reporting, as well as estimates of cost implication of each path of action being followed.

The *planning* part of the system is based on *missions* and *objectives*, rather than standard appropriation categories. This diminishes the relevance of

* D. N. Chorafas, *Risk Pricing*, Harriman House, London, 2010.

departmental boundaries and therefore, gives management a better view of operations. This is essential, as a major program is the province of all departments. *Programming* requires weighting, evaluating, and applying all of existing knowledge to the process targeted by PPB.

*Program budgeting*, for example, aligns financial outlays in terms of each major goal, in contrast to the functional or input requirements followed in the most corporate and governmental budgets. This sees to it that PPB's relevance to management is multifold, whether we talk of industrial firms, financial institutions, or the public administration. A fact finding PPB study focusing on future impact of current management decisions and resulting cost/benefit trade-offs among alternatives. Three structures can be distinguished as follows:

1. Across-the-board program grouping of fundamental objectives and associated organizational activities, into categories which contribute to specific end results
2. A multiyear program based upon decisions made to evaluate financial requirements; or, alternatively, proposals on major changes
3. Program updating procedures. To fulfill this mission PPB requires explicit information on decisions made about the allocation of resources and measures taken to position the company against changing market trends

Aside decisions made by management on resource allocation, one of the major sources of new proposals is program analysis itself, which may originate at any level in the organization. Its *accounting aspects* serve to record and present the financial effects of business transactions enabling to measure costs, benefits, and other financial results, with comparisons and ratios presented in graphic form, but backed up through analytical evidence if necessary. Basic questions PPB asks and attempts to answer are as follows:

- *On objectives*: How well are the aims of the organization fulfilled? Is what it is trying to achieve on target?
- *On alternatives*: What are the projected full costs of each alternative? Present normal costs? Long tail costs?
- *On effectiveness*: Which of the alternatives will be most effective, at a given cost, for achieving the desired objective? How does the cost compare with the benefits?

As these examples demonstrate, PPB helps to clarify a given mission, starting with input requirements. To successfully accomplish data collection, retrieval and reduction *goals* and *missions* must be stated precisely. *If* stated too broadly, *then* they are of little analytical value. In addition, particular attention must be paid to

- goals considered dominant because they influence spending decisions, and
- programs sharing the same goal which cross lines of executives authority and responsibility

The comparative advantage of PPB is that, in spite of practical problems, the very process of thinking about what one is doing, and for whom, adds a new dimension to management. This is further underlined by the factor of weighing alternatives. Setting up one way against another enables asking whether it is better to spend an additional sum for Subject A respectively reducing funds for Subject B.

Years of experience demonstrate that, to be effective, a PPB evaluation in an industrial company requires a careful attention to the way the job is formulated. Invariably, the orderly handling of details in projecting facts and figures about future operations will call for the implementation and observation of systematic standard costing procedures, along the examples given in Chapter 15.

In conclusion, the major function of PPB is to serve managerial requirements for dependable information thereby facilitating the attainment of objectives. It is concerned with the systematic evaluation of facts and figures about the operations of the enterprise; involves the polling of expert estimates through Delphi; utilizes procedures related to performance; aims at minimization of errors, fraud, and waste in carrying on the operations; and assists in a better documented preparation and administration of budgets, associating them to organizational units and responsibilities.

## 16.7 MANAGEMENT ACCOUNTING AND VIRTUAL FINANCIAL STATEMENTS

No matter what the books may be writing, *accounting* is not a science that obeys strictly *quantitative* terms though rules promulgated by state authorities (which change over time and from jurisdiction to jurisdiction) stipulate how accounting facts should be reported. Accounting is as well *qualitative* and *judgmental*. Who is to say

- what is the correct allowance for bad debts,
- the appropriate reserve which should be made for law suits, and
- the proper presentation of accounting results for management control reasons

These questions are particularly relevant in the context of *management accounting* whose mission is to inform the company executives about the finances of the firm in an analytical, factual, and well-documented manner. The mission of management accounting is internal to the enterprise, and it should not be confused with the regulatory quarterly and annual financial reporting (more on this later).

To a very large measure, for reasons stated in the preceding paragraphs, internal management accounting includes a great deal of judgmental interpretation. Even in regulatory financial reporting, in spite of rules that specify how accounting data should be shown at the end of the day the presentation of results is itself an art whose heterogeneity can reach significant proportions.

This lack of homogeneity practically present in all accounting systems brings in perspective the need for standards needed to reach a consensus, by presenting account information in a comprehensive manner. Standardization and homogeneity are welcome but they pose a series of challenges, which even within the same organization and for management accounting purposes typically include

- identifying the most significant interdepartmental and intersubsidiary standards issues requiring resolution,
- establishing appropriate priorities when tackling those issues, which sometimes derive from company culture,
- achieving general acceptance for the standards to be, and assuring they are observed in reporting, and
- maintaining support not only for the process of standards-setting but also for its general implementation throughout the enterprise

It should not come as a surprise that similar challenges exist as well with *regulatory financial reporting* and it is wise to align internal management accounting with standards and forms required by regulators. This way, executives receiving internal reports will not have to look at two incompatible modes of presentation. The only difference is that internal management accounting can take some freedoms that are not permitted with regulatory reporting.

Government-sponsored accounting standards organizations treat a whole range of issues in terms of standardization. In the United States, the Financial Accounting Standards Board (FASB) is constantly deliberating on norms connected with financial reporting models. It also devotes time to more narrow scope practices concerning some of the mechanics of the financial accounting profession itself.* Setting standards for the EU, and many other countries, is the remit of London-based International Accounting Standards Board (IASB).†

As in the case of technology standards set by the International Standards Organization (ISO) and the American Standards Institute (ANSI), financial and accounting standards provide basic references constituting a conceptual framework for resolving accounting issues. These help to establish reasonable bounds for judgment in

- preparing financial information, which is sound and useful, and
- increasing the recipients confidence in the contents, so that they make better informed decisions

The initiatives FASB and IASB take rest on the understanding that accounting standards are essential to the proper functioning of the economy. Decisions about the allocation of financial and other resources rely heavily on credible, concise, and comprehensive financial information—and this statement is also valid for internal financial reporting.

The difference is that while regulatory financial reporting is providing information to people and entities outside the business, namely shareholders, bondholders, bankers, regulators, and other parties, management accounting is concerned with the internal dissemination of accounting information.

---

* Issues concerning the public sector are addressed by the Government Accounting Standards Board (GASB), a twin organization to FASB which has been established in 1984.

† Which has elaborated IFRS. (D. N. Chorafas, *IFRS, Fair Value and Corporate Governance. Its Impact on Budgets, Balance Sheets and Management Accounts*, Butterworth-Heinemann, London and Boston, 2005).

A common ground between regulatory financial reporting and management accounting is the ever present need of improving the quality of reporting. Associated to this is the mission of better understanding by all recipients of each information element, as well as of the nature and purpose of what is contained in such reports.

Reliable information about the operations and financial position of the organizations has many uses, while incorrect or incomplete accounting and financial information is worse than useless; it is misleading. The usefulness of financial reporting can be improved by focusing on the following:

- Accuracy of reporting as its primary characteristic
- Timeliness of information so that it reflects actual conditions
- The ability to proceed with evaluations through comparability and consistency

To fulfill this mission the financial division and IT operations should promptly consider and correct deficiencies in data collection, analysis, and reporting—while adopting every significant development in technology. An example is the virtual balance sheet introduced in the late 1990s and promptly adopted by tier-1 organizations.

Companies using Paleolithic IT (mainframes, batch, and Cobol) typically take months to close their books. But using real time and expert systems by the late 1990s companies ahead of the curve were able to do so in 24 hours. This time lag has been further squeezed through *virtual balance sheets* featuring an accuracy of about 4 percent—which is enough for rapid response to management queries on assets and liabilities. Other improvements, too, have come along. Using prognosticator models management is offered a month ahead estimates on

- income
- expenses, and
- gross margins

Once established, the facility underpinning virtual balance sheets can be extended and financial reports can be personalized to different management levels. An individual product line manager can see exactly what the gross margins are on products under his or her watch—pinpointing those that are below expectations. This enables him to determine in a fly if below expectation conditions were caused by a market switch, delays, cost escaping control, discounting by competitors, or other reasons.

With such information on hand corrective action can be focused and immediate. It is therefore surprising that, whether addressed to senior management as virtual balance sheets, or to functional management as real time ad hoc reports, the "virtual close" of accounting data has not yet become commonplace.* Only the leaders have been able to shrink the time necessary to gain insight into the financials. The others live still in the Dark Ages.

* A most likely reason is that it cannot be supported by obsolete IT, and too many companies still feature legacy IT.

# Index

## C

## D



## E

## J



## K



## L

## Q

## T

## X



## Y



## Z